Serra · Zanarini   Complex Systems and Cognitive Processes

Roberto Serra   Gianni Zanarini

Complex Systems and Cognitive Processes

With 71 Figures

Springer-Verlag Berlin Heidelberg New York
London Paris Tokyo Hong Kong

Roberto Serra
DIDAEL
Via Lamarmora 3
I-20122 Milano, Italy

Gianni Zanarini
Physics Department, Bologna University
Via Irnerio 46
I-40126 Bologna, Italy

The *cover picture* shows a ganglion cell, part of the retina of the eye of a rabbit, enlarged about 5 000 diameters. The cell responds to the movement in one direction of an object in the field of view by sending electrical impulses to the brain, and ignores motion in the opposite direction. The cell was identified and stained by Frank R. Amthor, Clyde W. Oyster and Ellen S. Takahashi at the University of Alabama.

CR Subject Classification (1987): I.2, G.0, C.1, F.1

ISBN-13: 978-3-642-46680-9   e-ISBN-13: 978-3-642-46678-6
DOI: 10.1007/978-3-642-46678-6

Library of Congress Cataloging-in-Publication Data
Serra, Roberto, 1952–
Complex systems and cognitive processes / Roberto Serra, Gianni Zanarini.
p. cm. Includes bibliographical references.

1. Artificial intelligence. 2. Neural computers. I. Zanarini, Gianni, 1940–. II. Title.
Q335.S435   1990   006.3–dc20   89-26096   CIP

© Springer-Verlag Berlin Heidelberg 1990

Softcover reprint of the hardcover 1st edition 1990

The use of registered names, trademarks, etc. in this publication does not imply, even in the absence of a specific statement, that such names are exempt from the relevant protective laws and regulations and therefore free for general use.

Media conversion: EDV-Beratung Mattes, Heidelberg
2145/3140-543210 – Printed on acid-free paper

To
Francesca, Eleonora, Elisabetta,
Lorenzo and Gabriele

Preface

This volume describes our intellectual path from the physics of complex systems to the science of artificial cognitive systems. It was exciting to discover that many of the concepts and methods which succeed in describing the self-organizing phenomena of the physical world are relevant also for understanding cognitive processes. Several nonlinear physicists have felt the fascination of such discovery in recent years.

In this volume, we will limit our discussion to artificial cognitive systems, without attempting to model either the cognitive behaviour or the nervous structure of humans or animals. On the one hand, such artificial systems are important *per se*; on the other hand, it can be expected that their study will shed light on some general principles which are relevant also to biological cognitive systems.

The main purpose of this volume is to show that nonlinear dynamical systems have several properties which make them particularly attractive for reaching some of the goals of artificial intelligence.

The enthusiasm which was mentioned above must however be qualified by a critical consideration of the limitations of the dynamical systems approach. Understanding cognitive processes is a tremendous scientific challenge, and the achievements reached so far allow no single method to claim that it is the only valid one. In particular, the approach based upon nonlinear dynamical systems, which is our main topic, is still in an early stage of development.

The human brain evolved by adopting an "opportunistic" strategy, and artificial cognitive systems should in an analogous way evolve towards an integration of different paradigms, in particular towards a coupling of dynamical systems with classical AI techniques.

The structure of this book reflects these beliefs. The most successful and most thoroughly studied dynamical cognitive systems are connectionist models: therefore much attention is given to neural network models. Indeed the volume can also be used as an introductory textbook about connectionism.

However, the most attractive features of connectionist models are shared by a wider class of dynamical systems. In our view, emphasis should be placed upon these properties of dynamical systems rather than on the fact that the latter could be interpreted as networks of highly simplified neurons. Therefore, room is left also for dynamical cognitive systems different from neural networks.

Classifier systems are given much emphasis, since they can provide a link between the dynamical and the inferential approach to AI.

The presentation given here is by no means complete. The literature in this field is large and rapidly growing: the emphasis placed upon the different models reflects reasons of scientific interest, historical importance, personal taste and, as in every human affair, randomness. Our choice was to treat, first and foremost, some models which allowed us to illustrate, as clearly as possible, what we believe to be the most significant characteristics of the dynamical approach to artificial intelligence. We do not mean that the space given to the different models is a measure of their scientific or applicative importance.

Moreover, this volume is concerned essentially with ideas, methods and techniques. We did not feel it appropriate to include a detailed account of work on applications, since this would soon become old, while we hope that at least some of the ideas presented here may have a longer decay constant.

This volume is the result of the joint work by both of us; however, the major influence in Chaps. 1, 3 and 5 was by GZ, and in Chaps. 2, 4, 6 and 7, by RS.

We wish to acknowledge here the support of the University of Bologna, Enidata and Tema to our work. Such support was not only financial, but also scientific and cultural. Particular thanks are due to Vincenzo Gervasio, Francesco Zambon, Silvio Serbassi and Paolo Verrecchia.

The development of the ideas presented here has been made possible by the stimulating collaboration with some friends and colleagues: Mario Compiani, Daniele Montanari and Gianfranco Valastro. Most of the results presented here have been obtained by working with them on specific research projects.

The contribution of some bright students (Franco Fasano, Paolo Simonini and Luciana Malferrari) has also been very important.

We have also enjoyed the benefit of deep and fruitful discussions with Marco Vanneschi and Fabrizio Baiardi about the relationship between dynamical networks and parallel computation, with Tito Arecchi and Gianfranco Basti about the role of chaos in neural models, with Francoise Fogelman about layered feedforward networks and with Luc Steels about non-neural dynamical systems for AI and about genetic algorithms. On this latter topic, also fruitful discussions with John Holland and Heinz Muehlenbein are gratefully acknowledged. Rick Riolo kindly provided us with the CFS-C software package for classifier systems. We also benefited from a stimulating discussion with Edoardo Caianiello about the past and the future of neural networks.

Thanks are due also to Derek Jones, who carried out a careful English translation of our work.

We finally wish to express our gratitude to Hans Wössner and to his staff at Springer-Verlag for their interest and their support to our book, and for their friendly and careful editorial work.

Bologna, January 1990

Roberto Serra
Gianni Zanarini

Table of Contents

6   Dynamical Rule Based Systems

7   Problems and Prospects

1. Introductory Concepts

1.1 Complex Systems and Self-organization

Observing many natural phenomena (such as a flame, a smoke-ring, a procession of clouds) we are often surprised by their regularity, their organization, their dynamical order. The science of complex systems has undertaken the study of these aspects of simplicity which emerge from interactions amongst a myriad of elementary "objects". It has stepped forward to answer the questions which present themselves when we observe more closely, with greater attention, many systems with which we have become acquainted from daily experience.

This order, which we often observe over a certain space-time scale, how does it arise? From where do the individual elementary objects get the information necessary for them to conform to the global order? Furthermore, it is an order which relates to a completely different space-time scale from any meaningful one at the level of the individual elements themselves. Where does the design for this emergent order lie? From where is it controlled?

The answer provided by the science of complexity centres upon the development of the concept of "self-organization", which expresses precisely this possibility of highly organized behaviour even in the absence of a pre-ordained design (Haken, 1978; Nicolis and Prigogine, 1977; Serra et al., 1986).

As an example, in order to clarify the genesis of a self-organized situation in a physical system, we will refer to a particularly interesting system: the laser. To a first approximation, one can consider this system as being composed of a volume of "active material" placed between two mirrors and of a suitable external energy source. The atoms of the active material function as "oscillators": once promoted to an excited state, they can emit electromagnetic radiation of a characteristic frequency; they can also absorb incident radiation of the same frequency. Emission, in particular, can take place either spontaneously or through a kind of "resonance" with the incident radiation ("stimulated emission").

Usually, in response to an external excitation, a system of this type shows an "uncorrelated" emission, such as that giving rise to light emission in a normal electric lamp. However, under particular conditions, defined by the geometry of the system and by the intensity of the external radiation, the oscillators get synchronized and operate in a cooperative mode, supplying a highly coherent emission of radiation. In this case, the light waves produced by the cooperative behaviour of the individual atoms act as "messengers of a possible order" towards the individual atoms themselves, whose behaviour they condition.

In this example (although its illustration has been greatly simplified for reasons of brevity) we can see the development of a self-organization which is a consequence of the collective behaviour of a large number of atoms. This self-organization, however, can be recognized and is meaningful only over a space-time scale which is different from that of the atoms, being much more aggregate.

Even within physics, one can find many other self-organization phenomena, so many that we begin to think that we may be changing our way of looking at the world: we observe today, with great attention and wonder, ordered dynamical structures which in the past we took for granted, or considered as being not particularly interesting.

Let us recall another case: the thermo-hydrodynamic instabilities in a fluid close to its boiling point. It is well-known that, under the action of a thermal gradient, produced and maintained externally, convection currents develop within the fluid. As in the case of laser, under suitable experimental conditions, these currents take shape according to regular structures of a characteristic "beehive" form (Bénard cells), which indicate the presence of a high level of molecular cooperation (Haken, 1978).

One striking aspect of these physical examples is the fact that the establishment of a self-organized behaviour depends upon parameters with quite a meagre information content about the characteristics of the self-organization. The geometrical structure and the radiation intensity in a laser, like the physical dimensions and the thermal gradient in thermo-hydrodynamical instabilities, "know nothing" of the self-organization which emerges in the system, just as the atoms of the active material or the molecules of the fluid know nothing of it.

As we shall see in Chap. 3, this observation can be reformulated in mathematical terms, because the cooperative effects and the corresponding self-organization processes may be described by means of nonlinear equations which usually allow multiple asymptotic solutions; the variation of suitable parameters can alter the stability characteristics of the solutions, so inducing transitions from one to another.

At this point, however, a cautionary note is necessary. These allusions to the themes of self-organization might seem to suggest that, in all cases, a greater order and a greater simplicity emerge at the more aggregated levels. In many cases, however, on aggregate space-time scales, highly complex situations may be observed which can be characterized in terms of "deterministic chaos" (Arecchi, 1986; Serra et al., 1986). This theme will be taken up again in Chap. 3.

1.2 Self-organization in Artificial Systems

It might seem that the above considerations refer exclusively to natural systems. But, if we reflect for a moment, we realize that this is not true. Take, for example, the laser: it is a system which, while being based upon natural processes and showing an effect which in some cases may be found in nature, is normally built and operated according to a design. Can we then say that what is observed in a laser is an example of "designed self-organization"?

The laser lies in an intermediate position between natural and artificial systems. Its designer, in fact, can neither foresee nor control the behaviour of the single elements (in this case, the atoms of the active material). The design essentially regards the control parameters which affect the system's behaviour.

A completely artificial system is one which does not critically depend upon a material support. The best examples of completely artificial systems are abstract mathematical and logical systems. Even though an abstract system is made up of a large number of elements, every one of them can be defined a priori. Therefore, we must clarify the meaning of self-organization in artificial systems.

In order to enlarge upon this point, we will briefly examine a particular class of artificial systems, which will be dealt with in detail in Chap. 3: one-dimensional cellular automata (Wolfram, 1986). Consider a large number of binary elements (i.e. which can only take the values 0 or 1) and suppose that these elements are placed in a regular fashion along a line, each being connected to its two "nearest neighbours" (next left and next right). At any discrete point in time, the state of each element is defined by a function of the states of the element itself and of its two nearest neighbours at the preceding time-step. For simplicity, we will assume that this function is the same for all the elements. To avoid boundary problems, we also assume that the system closes upon itself to form a circle.

A detailed examination of the various types of automata and their classification is given in Chap. 3. For the moment, we will show that cellular automata, although extremely simple from the point of view of their "elementary laws", may exhibit unexpected and complex behaviours which can be considered as self-organized.

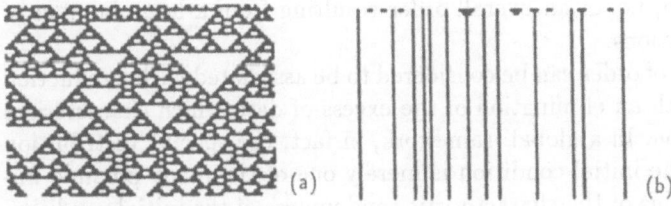

Fig. 1.1a,b. Two examples of one-dimensional cellular automata. Time evolution is from top to bottom of figure

To illustrate this statement, let us first of all consider the automaton in Fig. 1.1a, and follow its time evolution starting from randomly defined initial conditions. Black dots represent 1's, while white dots represent 0's. Time evolution is from the top to the bottom of the figure. As seen in the figure, the system shows a behaviour which is both ordered and predictable: in the space-time plane, in fact, we can see the appearance and the disappearance of triangular structures of various dimensions. From an undifferentiated initial state (that is, with the same probability of having one or the other of the two possible values at any position) the system moves on to a multiplicity of well-differentiated domains which evolve over time. Another ordered system is shown in Fig. 1.1b. In this case, after a short initial transient, the system behaves in an extremely regular manner, giving rise to a time-independent configuration. The detailed structure of this configuration (number of "lines" and distance between "lines") depends upon the initial conditions.

We can then say that, in these systems, a kind of spontaneous self-organization takes place, which brings them closer to the physical systems mentioned above. Their organization, in fact, does not derive directly from an external design: the design does not regard the whole system, but only the individual elements, whose interaction gives rise to an overall regularity (whether it be time-dependent, as in the case of the "triangles" in Fig. 1.1a, or time-independent, as in the case of the "lines" in Fig. 1.1b).

Self-organization, therefore, may arise not only from the interaction between unknown microscopic dynamics, but also from the practical impossibility of foreseeing all possible kinds of collective behaviour which can emerge from the interactions between microscopic elements.

This leads to an extremely important consequence: in a more radical way than in the case of natural systems, self-organization in abstract systems is not an observer-independent concept.

In order to describe the emergence of self-organized behaviours in cellular automata we have adopted a particular space-time scale: that is, we have considered to be meaningful a spatial dimension of the order of the whole network (unlike that which is meaningful for a single element, of the order of the distance between elements) and we have adopted an aggregate time scale (unlike that which is meaningful for the individual elements, whose memory is limited to a single time-step). It is from this particular point of view that the evolution of the automata in Fig. 1.1 may be described as an emergence of self-organization, i.e. of an overall order resulting from a large number of elementary interactions.

This emergence of order can be considered to be associated with a reduction in redundancy, with an elimination of the excess of data which characterizes the initial condition. In a global framework, in fact, the specific distribution of 1's and 0's in the initial condition is merely one of the many possible microscopic realizations of the characteristic randomness of the initial condition itself. In this sense, a detailed knowledge of the initial condition may be considered to be redundant.

On the other hand, if we observe the system by concentrating upon the space-time scales which are characteristic of the individual elements (that is, from a microscopic point of view), nothing important changes with time: neither the possible states, nor the number of nearest neighbours, nor the transition function. From this point of view, the state of the system at any time appears to be completely determined by the law of transformation and by the initial conditions. But this is not all: surprisingly, an observer adopting this microscopic point of view might even notice that the evolution of the system carries with it a reduction, an impoverishment of the initial information. The observer may explain this by the fact that the transformations produced by the transition functions in some cases compress the richness of the initial state into a repetitive homogeneity (as, for example, in the case shown in Fig. 1.1b).

But why does that which is seen from an overall viewpoint as a useless redundancy of the initial conditions now become a wealth of information which is destroyed over time? Precisely because the attention of a microscopic observer is not focussed upon the random nature of the initial distribution, but rather upon its detailed structure.

These two different approaches in considering the characteristics of the initial conditions are confirmed by posing the question in terms of the complexity of the algorithm necessary for defining them. Here, for complexity of an algorithm we mean, following Kolmogorov and Chaitin (Chaitin, 1975), the minimum length of the program necessary to carry out the required task. It can then immediately be seen that the algorithm required to obtain a string of random numbers is, from the overall viewpoint, quite simple, whereas, from the microscopic viewpoint, it is the most complex algorithm that one can imagine, because there is no way in which the corresponding program can be made shorter than the description of the string of numbers which constitutes the output (Atlan, 1987).

To summarize, it can be said that, from the macroscopic viewpoint only, that which emerges is meaningful, distinguishing itself from the insignificant redundancy of the microscopic level, whereas at the microscopic level, not only are the transition functions and the interconnections important, but also the detailed information about initial conditions.

Therefore, if we find ourselves in either of these two points of observation, that which is important for the other may lose its importance. These brief reflections confirm the relevance, the central role of the observer in the analysis of systems and, in particular, in the identification of their self-organization characteristics.

1.3 Cognitive Processes in Artificial Systems

We would now like to further extend our considerations about the centrality of the observer in the identification and in the description of self-organization processes. As we shall see, in fact, it is the adoption of a specific viewpoint which allows the recognition and the description of cognitive processes (Ceruti, 1986; Varela, 1986) in a system.

Let us first of all take the viewpoint of an observer external to the system, which studies its interaction with an environment which is, a priori, endowed with a meaning: the "design and control" point of view. Then one has to see if and how the system, at least partially, recognizes the meaning of the input.

In this light, for example, Fig. 1.1b may be interpreted in the sense that the system "responds" to the initial conditions with a behaviour (which can be detected externally through the time sequence shown in the figure) which "recognizes", in the initial conditions, particular spatial sequences of 1's and 0's.

On the other hand, by taking up a point of view "internal to the system" (however vague this expression may be in the case of systems which are not capable of self-consciousness), attention will be focussed upon its properties of autonomy, that is, upon its autonomous "creation of meaning" for the experience of the external environment: an experience which, in the case of the example in Fig. 1.1b, consists only of the initial conditions.

This change of viewpoint leads to consider the environmental influences as having, a priori, no meaning for the system. In other words, in this latter case it is not assumed that an "absolute" meaning is given for the external environment, but rather a creation of meaning is observed by the system itself (Varela, 1986).

The previous example of cellular automata does not allow a more detailed examination of the differences between the two approaches towards cognitive processes in artificial systems mentioned above. In fact, in order to observe an effective creation of meaning, a building of representations, a recognition of configurations, a learning from the external environment, it is necessary to consider systems which can change. We will take up these arguments again in the following chapters, where learning in artificial systems is discussed.

One can, however, speak of cognitive behaviour in artificial systems both from a "control" perspective and from an "internal" viewpoint. Thus, it is not a question of choosing once and for all between the two points of view which have been briefly outlined here. It would be much better to recognize their complementary nature, and to adopt one or the other according to the objectives set.

For example, it is clear that if cellular automata are to be studied as the first example of complex artificial systems capable of cognitive behaviours (such as the recognition and learning of "patterns" which are meaningful to an external observer) then the adoption of a "control-centred" approach is certainly an adequate one. If, on the other hand, the reference to cellular automata has the function of favouring the comprehension of concepts which are central to the description of biological systems (such as, for instance, the creation of meaning), then the adoption of a "control-centred" approach may turn out to be misleading and trivializing, and an "internal" perspective seems a more adequate one when our attention is directed towards the emergent cognitive dimensions.

1.4 Metaphors of the Cognitive Sciences

We have presented so far some examples and reflections which may help the reader to appreciate the richness and the fascination (scientific and epistemological, besides aesthetic) of the science of complex systems, and its potential application to the study of cognitive systems. At this point, however, some historical aspects are worth mentioning, in order to better appreciate the novelty and the effectiveness of the complex systems approach towards the study of cognitive processes, whilst succeeding, at the same time, to understand its roots. This theme will be taken up in more detail in the next chapter.

Cognitive processes are the object of research in various disciplines which, over the last 50 years, have experienced varied and continually changing interrelationships: neuroscience, psychology and information science (Parisi, 1989).

It is quite difficult to summarize the differences in the various approaches, without running the risk of being over-schematic. It may be stated, however, that neuroscience studies the neurocerebral system as a physical apparatus which shows computational properties (in the etymological meaning of analysis and evaluation of different pieces of information): input recognition, reasoning, learning, etc. Neuroscience essentially focusses upon the microscopic level, attempting to explain the working of the brain at an overall level by reducing it to elementary processes.

Psychology, on the other hand, deals more with the mind than with the brain, i.e., it deals with cognitive behaviours manifested by living organisms having a neuro-cerebral apparatus. Even when the prevalence of a reductionist approach has emphasized the expectation of a definitive explanation of the working of the mind on the basis of the underlying biochemical processes, psychology has always maintained an approach centred on high levels of aggregation.

Information science, unlike the previous ones, is not characterized by a precise option about the level of aggregation for the study of its own objects. On the contrary, it is possible (at least schematically) to distinguish in its history various phases corresponding to different approaches, and therefore also to different relationships with the other disciplines cited above. Moreover, information science, precisely because of its greater variety of different viewpoints, may constitute an important reference for a better articulation between aggregate and microscopic approaches in the cognitive sciences as a whole.

When information science, under the name of cybernetics, began the formal study of mental functions, it was characterized by a tendentially microscopic approach. It could not, in fact, be otherwise, as one of the objectives of cybernetics was that of building machines capable of cognitive behaviours: its aim was not only to explain, but to explain operationally, achieving high-level functions starting from very simple elementary cognitive mechanisms. This particular orientation of cybernetics, as may be expected, favoured the development of close links with neuroscience.

For various reasons (some of which will be examined below), this approach had limited success, and the informatics community thus created the expression "artificial intelligence" to indicate a new direction of information science: the high level approach, in which cognitive functions began to be studied independently of their physical implementation. Increasingly, the "artificial minds" moved away from the "electronic brains" which supported them, rendering themselves autonomous. This transition, abandoning the hypothesis that artificial cognitive systems could reap advantage and inspiration from the study of the physical characteristics of the brain, led information science further away from neuroscience but closer to psychology.

This allows the understanding of the birth of a whole new line of psychological research (cognitive psychology) inspired by the analogy between symbolic computing by the mind and the operation of computers (in their high-level characteristics, and no longer in their micro-organization). In particular, within the field of cognitive psychology, the drafting of computer programs in high level languages has increasingly taken on the function of simulating mental processes, although with the awareness that complete reduction is not possible.

Some of the problems currently facing artificial intelligence will be examined in the following chapters, together with its undeniable successes. We will limit ourselves here to recalling the difficulties and the costs associated with the storage of knowledge in the form of rules, the problem of managing contradictions and uncertainties, and the fragility of artificial intelligence systems. These difficulties force to reflect more deeply upon the convenience, and even the possibility, of completely separating the "artificial mind" from the "physical machine" supporting it.

It is true, in fact, that many high level cognitive functions can be implemented upon a multiplicity of different substrates. But it is also true that a particular organization of a substrate (i.e., its microscopic structure) may give rise, through self-organization processes, to high level primitives whose implementation would, perhaps, otherwise not even have been attempted.

The development of a science of complex systems, which stresses the importance of self-organization processes, can give a decisive contribution in overcoming many of the difficulties faced by artificial intelligence, showing the potentialities of "connectionist" architectures, that is of structures made up of an enormous number of identical and mutually interconnected elements. Moreover, also the developments of neuroscience have contributed to the re-launching of the cybernetic approach.

At the same time, cognitive psychology has come up against the increasingly evident limits of the approach adopted. The "commonsense reasoning", the spontaneous generalizations, the learning by examples instead of by rules (to cite only a few points), that is, the simplest cognitive capabilities of a young child, appear to be rather difficult to include within the framework of the "metaphor" of the symbolic processing of information which inspires cognitive psychology. Thus, even psychology has been compelled to search for new reference models which could, at least, complement the preceding ones where these latter fail. The new "neurally inspired" (not purely "mind inspired") information science increasingly constitutes, in this sense, a promising "workshop" for the construction of models capable of interconnecting the various levels of analysis and of emulating cognitive behaviour.

In a sense, a historical circle has been closed, and information science is today "revisiting" the cultural environment in which it was born. But its history is not one of lurchings and waverings: it is rather the history of a relationship between a complex system (the mind-brain system) and a complex task (the development of artificial systems with cognitive capabilities).

2. The Dynamical Systems Approach to Artificial Intelligence

2.1 Introduction

In the preceding chapter we stressed the importance of the science of complex systems for understanding and simulating cognitive processes and we outlined the relationships between the various disciplines which, from different points of view, deal with cognitive processes.

In particular, we discussed the relevance, in this context, of the spontaneous emergence of global properties, which can be shown to derive from the interaction of simpler elements, but which can not be adequately described using the same language used for describing the microscopic level. The emergence of such properties is a fundamental characteristic of complex systems although, from the "control-oriented" perspective, they can play an unpleasant and undesired role.

In any case, an artificial system (abstract machine) capable of cognitive behaviour must necessarily rely on some form of spontaneous organization of relatively simple elements. The complete and exhaustive external design of such a machine is, in fact, too complex to be considered as a realistic prospect.

But this is not only a practical problem: indeed, we wouldn't consider "intelligent", in any sense, any machine incapable of such a form of organization. Suppose to try to construct a machine – say, a computer program – for classifying a given set of patterns. Let us call P the set of possible input patterns to be classified, while the set of classes, the outputs, will be called C.

In principle, a possible solution to this problem consists in a lookup table which associates to every element in P the corresponding element in C (although this solution would be practically impossible in almost every interesting case). The important point, however, is that even if such solution would be possible, we would not regard such a system as an "intelligent" one, but rather as a conventional data base.

On the other hand, an external observer, totally unaware of the internal working of the machine, might consider it as an impressive demonstration of the power of artificial "intelligence".

We are faced here with a fundamental problem in the study of complex systems, namely that of the choice of the appropriate viewpoint. In this volume we shall most often take the viewpoint of the designer of an "intelligent machine", who is perfectly aware of its functioning at the "microscopic" level, and is interested in achieving good performances at a global level.

In this chapter, we initiate a comparison, which will run through the whole volume, between two basically different methods for achieving this goal of "making simple parts work together properly". One is the chaining of inferences, ruled by a central control program, to which classical artificial intelligence (AI) refers, while the other is dynamical self-organization, which lies at the heart of the "neural" or "connectionist" approach.

Let us also remark that some restrictive definitions of artificial intelligence have been given, like e.g. the recent proposal by Kodratoff (1989) who identifies it with "the science of explanations". If such definitions were adopted, then it would be meaningless to discuss of a "dynamical systems approach" to AI. However, we will refer to the widespread use of the term as meaning "designing intelligent systems, that is systems that exhibit the characteristic we associate with intelligence in human behavior" (Barr and Feigenbaum, 1981).

In Sect. 2.2 we will describe the general features of the dynamical systems approach to artificial intelligence.

Actually, this approach is based upon many old ideas of cybernetics, which were however gradually abandoned in favour of the symbolic approach of classical AI. In order to avoid old mistakes and to overcome the past limitations, it is necessary to analyze the reasons of the failure of the old dynamical approach, and to understand what has changed since then. This is done in Sect. 2.3.

Sect. 2.4 is devoted to a comparison between the classical and the dynamical approaches to AI. We believe that the two paradigms – inferential and dynamical – should not be regarded as opposites but rather as complementary. They can be adopted simultaneously, within the framework of a cognitive artificial system, to carry out different interacting functions. Moreover, artificial systems can be designed which intriguingly mix elements from both approaches: an example is that of classifier systems, to which an entire chapter is devoted.

However, in order to better illustrate the distinguishing features of the dynamical approach, we will emphasize, in this initial comparison, the differences between the two approaches, postponing until the final chapter the discussion of their complementary nature and possible synergies.

2.2 Dynamical Systems, Attractors and Meaning

We will now discuss the relationship between the dynamics of complex systems, which will be studied in detail in the following chapter, and cognitive processes. We are interested above all in the construction and in the understanding of artificial systems: thus the term *cognitive processes* is used to indicate complex operations which recall, in a general sense, those characteristic of human mental activity without, however, attempting to reproduce the latter with precision.

How can the evolution of dynamical systems be associated with the execution of cognitive tasks? A dynamical system, starting from an initial condition, passes through a succession of states. If the initial state is associated with a certain description of the world, the successive states consist of transformations of this description, and may be considered as a "processing" of the initial information.

Let us consider, for example, the classical problem of alphanumeric character recognition, using a dynamical system composed of N interacting boolean variables, which can be represented on a two-dimensional grid. In this manner, each letter can be represented by associating a dark point to all active variables (with a value "1"), and a blank space to those which take a null value (Fig. 2.1).

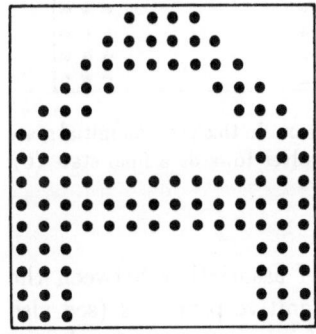

Fig. 2.1. A pattern can be associated to a dynamical system. In this case, the letter "A" is represented by boolean variables, associating a value "1" to black points and a value "0" to white points

Let us now assume that we can shape our system so that the representations of the letters of the alphabet become "fixed points", each with its own "basin of attraction". The concepts of "fixed point" and "basin of attraction" will be examined more closely in Chap. 3. For the present it is sufficient to briefly recall some definitions.

An *attractor* is a state towards which the system may evolve, starting from certain initial conditions. The basin of attraction of an attractor is precisely the set of initial conditions which give rise to an evolution terminating in that attractor. An attractor may consist of a unique state, in which case it is referred to as a *fixed point* or a *stationary state*, or of a periodic succession of states (*limit cycle*), or it may have a more complex structure, as will be seen in Chap. 3.

For the moment we will not discuss the manner in which it is possible to obtain the desired result, i.e. associating the fixed points to the desired states: this topic will be treated in detail in the following chapters, when various algorithms proposed for this end will be examined more closely. Thus let us assume that we have reached our aim, i.e., that we have associated the various letters of the alphabet to attractors of the dynamical system.

Suppose now that we have to examine a letter which is incomplete or ruined by superimposed random noise. The initial condition will thus be of the type shown in Fig. 2.2. If it falls within the basin of attraction of the "right" letter – that is, if the latter is still recognizable notwithstanding the noise or the mutilation – then the system will evolve until it reaches the configuration corresponding to the original form of the letter (Fig. 2.2). In this case, it may be said that the recognition has been correctly carried out.

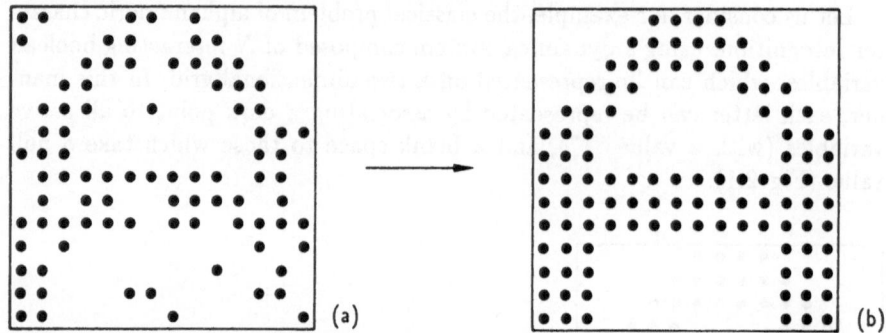

Fig. 2.2a,b. Dynamical evolution may correspond to recognition. In this case, an initial state (a), which represents a noisy picture of the letter "A", evolves towards a final state (b), which corresponds to a correct picture of that letter

This example illustrates the general scheme of association between the dynamical behaviour of complex systems and cognitive processes (see also Smolensky, 1988): certain "states of the world" of interest are mapped onto the dynamical system. They can, for example, be represented by the initial state of the variables, as in the case described in Fig. 2.1 and 2.2, or by the initial state of a properly chosen subset of the system variables. This, however, is not the only possibility: other alternatives consist in representing the states of interest using the values of the model parameters, or using external inputs to the system. It is also possible to use various combinations of the choices described above (initial conditions, parameters, external inputs).

The law of dynamical evolution will then generate a succession of states of the system, which correspond to transformations (interpretations) of the initial conditions. The situation is particularly simplified, and rendered more manageable, if the transients are ignored and only the asymptotic states are considered. The latter then represent the "conclusion" which the system reaches, starting from the initial conditions and from the given values of the parameters.

This correspondence is intuitively simple. In order to illustrate it further, however, let us consider a case where the information is represented by the values of the parameters, instead of by the initial conditions. The simplest case is that of a unidimensional system such as, for example, the logistics map which will be studied in Sect. 3.3. This is described by the equation:

$$x(t+1) = Ax(t)(1 - x(t)) \tag{2.1}$$

Let us assume that the information on the state of the external environment is represented by the value of the parameter A. It can be shown that this simple equation gives rise to different asymptotic behaviours, corresponding to the different values of the parameter A. In particular, if $A < 1$ the origin is the only stable fixed point, whereas if $1 < A < 3$ the only stable asymptotic state is given by the point $1 - 1/A$. If, however, $A > 3$, we encounter a period 2 limit cycle and, on further increasing the value of the parameter, the so called "subharmonic cascade" of cycles is found with doubling periods, until a chaotic regime is found. The relevant point for this discussion is that the asymptotic state of the system provides information about the value of A and, in particular, classifies it as belonging to a particular interval of values. The system, in a certain sense, acts both as a "recognizer" and as a "classifier" of the value of the parameter A.

This example has been mentioned to further illustrate the nature of the implementation of artificial cognitive processes using dynamical systems, showing a case in which the information is coded in the parameters instead of in the initial conditions, but above all it will serve us to open a critical discourse upon the considerations made so far.

These observations highlight a very important point: any dynamical system, no matter how simple, is capable of carrying out operations which may be considered by an external observer as very limited forms of recognition or classification. However, although the example of character recognition rightly belongs to the universe of cognitive processes, it is inevitable that the logistic map, as a cognitive system, appears to be rather disappointing. In order to avoid abusing the term "cognitive dynamical systems", therefore, it is a good rule to limit its application to those dynamical systems which can be shown to be capable of executing non-trivial cognitive tasks – according to some criterion such as, for example, the subjective evaluation of the apparent difficulty of the task, or the difficulties involved in programming a computer to carry it out, etc. In any case, the border between "cognitive" dynamical systems and the others is all but well-defined.

A non-trivial cognitive task requires the capability of internally representing a large amount of information, and to carry out diversified classifications. Therefore, dynamical cognitive systems need to have several variables and numerous possible asymptotic states. It is thus clear that we must deal with systems with a large number of degrees of freedom interacting in a nonlinear manner.

To highlight the interaction between the different variables, we will use the term "dynamical networks", imagining the variables as being associated with the nodes of a network, and connecting with arcs the variables which are effectively coupled through the equations. Later, many examples of networks of this kind will be met.

Dynamical networks can be continuous or discrete with respect to time; for reasons related essentially to the facility of implementing computer models the latter alternative is often preferred. Thus we will mainly consider discrete dynamical networks, which can be described by the general equation:

$$x_i(t+1) = f_i(x(t), P^{(i)}(t), E^{(i)}(t)) \qquad i = 1, \ldots, N \qquad (2.2)$$

with x representing the N dimensional vector whose components are the systems variables, $P^{(i)}$ representing a vector of parameters and $E^{(i)}$ possible external inputs (which could be stochastic). Obviously, this expression is of too general a nature to be of practical use. In the following chapters specific forms of this equation will be examined and their properties discussed.

Generally, we will deal with homogeneous networks, where the form of the function f is site-dependent only through the parameters $P^{(i)}$ and the external inputs $E^{(i)}$. We will also often consider examples lacking an external input to the system. In these cases the form of the evolution equation will then be:

$$x_i(t+1) = f(x(t), P^{(i)}) \qquad (2.3)$$

In principle, also non-Markovian systems may be considered, where $x(t+1)$ does not depend only upon $x(t)$ but also upon the values of preceding instants in time. However, all of these cases may be included in the formalism of Eqs. (2.2) and (2.3) simply by adding variables. For example, if the N-dimensional vector $x(t+1)$ depends both upon $x(t)$ and $x(t-1)$, it is possible to redefine the state of the system as a $2N$-dimensional vector $z(t)$, whose first N components are those of $x(t)$, while the successive N are those of $x(t-1)$. In this manner, $z(t+1)$ only depends upon $z(t)$.

A particularly important aspect relating to the dynamical systems approach to cognitive processes is learning. Actually, the capability of adapting to a changing environment, and of learning from past experience is a key property of cognitive systems.

A dynamical network, once the form and the values of the parameters which appear in Eq. (2.1) have been fixed, is capable of recognitions and classifications which are defined once and for all: the equations define its "cognitive repertoire". Methods are therefore required to model the phase space as a function of the specific requirements: for example, in the case of character recognition (Figs. 2.1 and 2.2), the various letters need to be associated with attractive fixed points. The network must be able, after presentations of the different letters, to shape its own phase space accordingly. If, for a certain class of models, the form of the equations is pre-defined, then learning consists of identifying an adequate set of parameter values. As will be seen in Chap. 3, it is possible in this way to profoundly modify the attractors of the dynamical system and their stability properties.

The modification of the parameters takes place according to a "learning algorithm", which depends upon the models adopted. As we shall see, in some models, like for example those which back propagate an error signal, this modification takes the form of gradual learning: there is a "teacher" which compares the final state (the response) of the system with its own criterion of adequacy, providing a feedback signal which is used to induce suitable modifications of the network.

By indicating the system adjustable parameters with W (they are a subset of the set of parameters P), the learning equation takes in this case the general form:

$$W(t+1) = g(x(t), P(t), L(x(t)))\tag{2.4}$$

where $L(x(t))$ denotes the teacher's evaluation of the system "answer" at time t. It is to be noted, however, that the time scale adopted here differs from that of the evolution Eqs. (2.2) and (2.3). Indeed, a single time step in the training phase requires a complete relaxation before the weights are adaptively readjusted.

In other learning schemes (e.g., the Hopfield model) there is "one-shot learning", and the external environment is directly coupled with the learning system. This means that the modifications of the parameters are a function of the patterns to be learnt, and do not depend upon the system's response to the presentation of these patterns.

In any case, dynamical learning systems for artificial intelligence show the interaction between two different dynamics, which in general operate on different times scales: the evolution dynamics of Eqs. (2.2) and (2.3), and the learning dynamics of Eq. (2.4). In several models, learning takes place in a certain phase of the system life, which is separate from the operational phase: in this latter phase, learning is turned off. Although this separation may be convenient for several applications, it is not always necessary, and there are models where it is not required (e.g., classifier systems). It is also clear that human learning is somehow continuously active.

The preceding considerations show the profound relationship existing between the dynamical approach to learning and the theory of adaptive systems. In the above exposition we have implicitly adopted the "control point of view": the task was pre-defined, the "meaning" attributed from the exterior.

From the "internal" point of view we can consider the adaptive modification of the parameters of the dynamical system as a particular form of interaction with the environment: the system is not isolated and modifies its own parameters (and thus its own phase space) due to this coupling with the exterior.

Whatever the learning mechanisms, there is in any case a flow of information between the outside world and the cognitive system. The process of learning in dynamical networks thus belongs to the class of self-organization processes in non isolated systems, several examples of such self-organization processes being found also in physical systems.

2.3 Neural Networks

So far, we have outlined some of the characteristics of "dynamical" cognitive systems, while we have not yet analyzed the inferential paradigm which, as is well known, is the one which has been most widely studied by the scientific community. It is interesting to point out that both of these conceptual references have been present from the beginnings of AI, and it is important to analyze the reasons lying behind the gradual prevailing of the logic oriented point of view. Only in the 80's, in fact, has there been a revived interest in studying dynamical networks.

This section is dedicated to a brief analysis of the evolution of the dynamical systems approach. It does not aim to provide an exhaustive review of the history of studies on dynamical networks, but rather to outline the reasons for their decline in popularity in the late 60's and for their recent revival.

In the period following the end of the second world war, the introduction of the first computers and, more generally, the development of very sophisticated machines brought the concepts of *organization* and *system* to the centre of scientific attention. The main goals of what became known as the "cybernetics movement" were twofold: on the one hand there was a need to develop a theory of complex machines and of their organization, to continually improve their design and utilization. On the other hand, there was the hope that the study of complex machines could contribute to the development of a theory of "living machines", and thus to an understanding of the organization of life itself.

These different motivations often co-existed in the same person. For example, von Neumann, the inventor of the architecture of the first computers – and thus a "designer" of complex machines – also initiated the studies on cellular automata, with the aim of understanding what were the necessary conditions for a system to possess the ability to reproduce itself.

One particularly ambitious subject, in this scientific environment, was the study of the human mind. The question was thus raised about the possibility that the same organizational principles which operate in machines and in biological systems could be at the root of understanding the functioning of the mind, and about the possibility of building "electronic brains" with capabilities similar to those of the human brain.

The computer itself was sometimes called "electronic brain" and the problems about the possibility that this machine could exhibit intelligent behaviour were greatly felt. For the first time, in fact, a machine capable of carrying out complicated "intellectual" operations became available – the first machine which constituted a "mirror of the mind".

From this background, two fundamentally different lines of research developed: the neural network approach and the symbolic approach. The former found its inspiration in the knowledge, albeit limited, about the architecture of the nervous system, while the latter operated at a higher level of abstraction. Slightly simplifying the two positions, it may be stated that the neural network approach was centred upon the study of the human brain and its emergent properties, while the symbolic approach was centred upon the study of the mind, independently of the structure and functioning of its physical support.

Direct modelling of the nervous system met with two major difficulties: the inadequacy of the computer resources and the ignorance about many essential aspects of the functioning of the system to be simulated. In this situation, the first research workers oriented themselves in a direction which proved to be extremely fertile in many branches of science, i.e. towards the construction of abstract models, extremely simplified, of the nervous system, and towards the study of their properties.

The first abstract model of neural networks was proposed by McCulloch and Pitts (1943). Basing themselves upon the knowledge of that period, they considered the neuron as a binary device, which can be assimilated to a logic unit which computes a logical function of its inputs.

The neuron of McCulloch and Pitts may thus be found in only one of two possible states {0, 1}. It may receive inputs from exciting synapses which all have the same value. If the sum of the inputs exceeds a certain threshold, the neuron is activated, otherwise it is not. Inhibitory synapses are also present, which have an absolute nature: in the presence of a signal deriving from an inhibitory synapse, the neuron does not fire, independently of the value of the excitation.

One can easily verify that a neuron of this type computes logical functions of its inputs. Usually, a high value is associated with "true" and a low one with "false". Let us assume that the exciting synapses have unitary intensity, and consider for example a neuron X with two inputs, which we will call A and B. The excitation function may thus take only three possible values: 0 (if the neurons A and B are inactive), 1, if only one is active, and 2 if they are both active. By varying the threshold of the neuron X, it is possible to obtain various logical functions: if the value is 1/2, we obtain A OR B; for 3/2, A AND B. If X had three inputs, with a threshold of 3/2 and the synapse from C were inhibitory, we would obtain A AND B AND (NOT C), etc.

McCulloch and Pitts succeeded in demonstrating that a network of formal neurons of this type could compute any finite logical expression. This result had a profound influence, since it showed for the first time that a network of extremely simple elements possessed an enormous computing power, a power which derived precisely from the presence of numerous elements and from their interactions.

However, the problem of building an intelligent "neural machine" is still a long way from being solved. In fact, the size and the complexity of the network make it difficult to imagine a technique for detailed designing of each of its component parts; besides, many interesting cognitive tasks can not find a "natural" expression in the predicate calculus proposed by McCulloch and Pitts. This is also a major difficulty in classical AI.

In order to make a neural network capable of complex tasks, it is necessary to find a self-organization mechanism which allows the network to "build itself", in relation to the task to be executed. We thus have to face the problem of learning: as mentioned in the previous section, the dynamical systems approach to AI leads to the association of learning with adaptive modification of the parameters. In the case of neural networks, these parameters can be naturally subdivided into two classes, those relating to a single neuron and those relating to the connections. If we set the neuron model – for example, that of McCulloch and Pitts – learning must then be associated with the modification of the synaptic connections.

It is thus necessary to find a method for changing the connection strengths according to the requirements of the task to be carried out. Also in this case, the study of artificial neural systems found its inspiration in the knowledge and hypotheses developed in the neurophysiological field.

The fact that synaptic connections vary from individual to individual, and that they vary over the course of time (Changeux, 1983), suggests that they are involved in the process of individual learning, and therefore, in particular, in the process of memorization. In which way this can occur is not yet completely understood. However, a hypothesis proposed 40 years ago by Donald Hebb (1949) is particularly elegant. It has had a great influence on the development of neural network models and has been confirmed, at least partially, by neurophysiological research.

To understand Hebb's idea, let us take a synapse which connects two neurons in an excitatory manner. Then, according to his own words, "when the axon of cell A is near enough to excite a cell B and repeatedly or persistently takes part in firing it, some growth process or metabolic change takes place in one or both cells such that A's efficiency, as one of the cells firing B, is increased" (Hebb, 1949). The idea is simple and elegant: synapses do not all have the same intensity, but modify themselves in order to favour the repetition of firing patterns which have frequently occurred in the past.

Hebb also introduced the concept of cell assemblies: according to this hypothesis, the cooperative nature of synaptic modifications is such as to induce the formation of subsets made up of cells which are mutually activated. A single cell may belong to several assemblies, and several assemblies may be active simultaneously, corresponding to "complex thoughts". The information would then be contained in these patterns of collective excitation, and would therefore possess a distributed nature. This proposal, moreover, provided an explanation for the discoveries of Lashley (1950) regarding the impossibility of localizing memories in the cerebral cortex.

Hebb presented his ideas in these general terms without stating them mathematically. Thus various mathematical models are possible, all inspired by the same qualitative hypothesis. In particular, Rochester et al. (1956) showed by computer simulation that a model of this kind, with some adjustments and modifications, effectively exhibits self-organization processes which give rise to the formation of cell assemblies.

This work is of interest because it also constitutes one of the first examples of the application of computers to the simulation of neural networks. Today, the interaction between theory, experiment and simulation is common practice in many sciences, but this was certainly not the case in the 50's.

A model of neural dynamics, introduced by Caianiello (1961), emphasizes the fact that neural networks are truly dynamical systems. The model was based on the separation between a set of "neuronic" equations, describing the time evolution of the neural activity (analogous to the evolution equations of the previous section, Eqs. (2.2)), and a set of "mnemonic" equations, describing permanent or semi-permanent variations in the coupling coefficients (thus playing a role analogous to the learning equations of Sect. 2.2). The model dealt with boolean neurons and took explicitly into account the delay introduced by the propagation of the neuronic discharge; the learning rule was of the hebbian type, with modifications due to the delay term. Recently, an anlytical solution of the neuronic equations was given (Caianiello, 1986)

Another approach to neural networks was based on linear models of the response of a neuron, which was supposed proportional to its input stimulus (Kohonen, 1972 and 1984, Anderson, 1972). The hypothesis of a neuronic linear transfer function allowed the use of techniques from linear algebra, with a good understanding of the model capabilities and limitations.

The output of the network is the matrix multiplication of the matrix of synaptic connections times the input vector. These models can be applied as associative memories and patterns associators, by defining the synaptic matrix as the outer product of the input and output vectors: in this way the synaptic matrix is the projection operator onto the linear space spanned by the stored patterns. This choice is also related to the Hebb hypothesis, since it gives high values to those coefficients which couple neurons which are often in the same state in the set of patterns.

Although these are static systems, it is important to consider their relationship with dynamical networks of the kind considered in this book. The discussion of the relationship between static descriptions based on projection operators and dynamic descriptions will be carried on in Sect. 7.3. Other interesting works of historical importance, related to "static" network models, include those of Steinbuch (1961), Pribram (1971), Gabor (1969), Willshaw et al. (1969). It should be recalled that also static networks can learn sequences of pattern associations, instead of single ones (Amari, 1972, Bottini, 1980).

Besides Hebb's hypothesis, another learning rule which has experienced significant popularity, and which has been recently re-vitalized by new discoveries, is the so-called "perceptron" learning rule (Rosenblatt, 1958). This rule is closely related to another learning algorithm proposed at about that time, Adaline (Widrow and Hoff, 1961). Since the perceptron has been more popular in the cognitive literature (Kohonen, 1984), we will briefly discuss some of its features.

Hebb's rule reinforces the correlations between already correlated neurons, and thus favours the formation of organized structures. The response from the environment, however, is lacking, i.e., it is not possible to state whether a correlation is useful or disadvantageous to the system.

The perceptron, on the other hand, interacts with a "teacher" which evaluates the response from the system and provides a feedback signal about its adequacy. In this case also, learning means modifying the connections, but here the modification is guided by an error signal. The process of learning is thus based upon the presentation of a given input to the system, the calculation of the corresponding output and the comparison with the exact response, which is known by the teacher.

The rule for changing the adaptable weights (perceptron learning rule) will be analyzed in detail in Chap. 5, and so will not be discussed here. It does however have a simple intuitive basis: one modifies only those weights which lead to erroneous output values, in such a way to force them towards a correct classification. The perceptron is a structured system, in which the input layer is distinct from the output layer and from the intermediate layers, whose function will be clarified later.

Perceptrons aroused much interest, also because Rosenblatt demonstrated that the learning algorithm was capable, in a finite number of steps, of finding a set of values for the synaptic weights which led to the correct classification, as long as such a set of values existed. Some practical demonstrations and the actual realization of a perceptron (using the hardware available at that time) led to a growth of interest.

As often happens in AI, the interest and hopes were, however, the prelude to successive disappointments. The problem is that, in many cases, the weights which lead to the correct classification can only be determined in the presence of an internal architecture designed ad hoc. Minsky and Papert (1969) discussed these aspects in detail in their well-known book "Perceptrons", which highlighted the structural problems connected with the learning ability of perceptrons.

During the same period there was a drastic decline in interest, and in financing, for neural network studies. According to some authors (McClelland and Rumelhart, 1986), "Perceptrons" played a decisive role in this sense. In any case, a single book can not be responsible for the decline of a scientific sector. The reasons are rather to be found in the limited success and in the disappointments of the dynamical systems approach.

Evidently, the understanding of thought and mental processes is not only an intellectual challenge of extraordinary interest, but also takes on profound emotional connotations. Every investigation on thinking machines is also, in part, an investigation on the thinking process and on the human being itself. The emotional implications can be seen, for example, in the appearance of excessive and unrealistic expectations: the entire history of AI is spangled with retracted optimistic forecasts (see e.g. McCorduck, 1979).

The neural network field is not different: the diffidence of many towards a machine-oriented approach to cognitive processes was added to the disappointment of ex-enthusiasts, thus giving way to a sharp decline in research in this field.

Alongside reasons of this kind, which belong to the sociology of science, there are more specifically technical ones. One is certainly a technological fact: the dynamical equations of neural networks can not be solved analytically, and it is necessary to resort to computer simulations or direct hardware implementation. Both the computing resources available and the level of electronic technology during that period were, however, overwhelmingly insufficient for adequately treating dynamical systems which were so onerous from a computational point of view.

A more profound reason lies, however, in the scanty understanding of nonlinear dynamics. That period witnessed the development of computing tools for linear systems with many degrees of freedom, and local linearization represented the main route towards facing nonlinear problems. The study of nonlinear dynamics was at that time confined to limited sectors of mathematical physics.

Let us now briefly examine the main reasons behind the recent revival of interest in the field of neural networks.

It was in the 70's and, with greater emphasis, in the 80's that the physics of complex systems placed strongly nonlinear systems at the centre of scientific interest. Therefore, while twenty years ago high dimensional nonlinear systems like neural networks were still too remote from major scientific interest and from the theoretical instruments then available, today they constitute a frontier for the science of complex systems.

The Hopfield model and statistical mechanics are particularly clear examples. As will be discussed in more detail in Chap. 4, the Hopfield model (Hopfield, 1982) is a network of Boolean neurons with Hebb-type learning. The two states of the neurons can be associated with the two possible values of electron spin in an external magnetic field. Based upon this correspondence, the Hopfield model presents a very close relationship with disordered physical systems, the so-called spin glasses (Mezard et al., 1987) which are at the centre of current research in statistical mechanics, since they show random and competitive interactions which give rise to a phase space with a particularly complex structure. The discovery of this relationship has involved an entire scientific community, that concerned with statistical mechanics, in the development of the theory of neural networks, significantly contributing to the present renaissance of these studies.

Similar interests involve a number of researchers in nonlinear dynamics who are becoming interested in the problems of neural networks. Indeed, although the mathematical theory of high dimensional nonlinear dynamical systems is not so mature as the corresponding theory of low dimensional systems, it now seems to be an interesting field worthy of exploration, with the aid of conceptual and computing tools developed over the last few years.

Another stimulus towards increasing interest in neural networks derives from progress made in neurophysiology, which has developed very precise and sophisticated analytical instruments. Interaction between this experimental science and theoretical models may turn out to be extremely fruitful.

The inadequacy of computing resources, while still present, is now greatly reduced with respect to the past: currently it is quite common to find workstations with a processing speed which, in the 60's, was typical of supercomputers. The possibility of running experiments and computer simulations of models is now much greater. It should be recalled, however, that the simulation of neural networks (strongly coupled nonlinear systems with many degrees of freedom) is still burdensome for computers based upon a traditional architecture.

Another significant stimulus towards the study of neural networks comes indeed from the developments in electronic technology and in particular from very large scale integration (VLSI) techniques. This allows the construction, at limited cost, of microcircuits which directly implement neural network models (Denker, 1986; Graf et al., 1986; Sivilotti et al., 1987). An alternative can be found in optical implementations (Denker, 1986; Psaltis et al., 1988; Anderson and Erie, 1987; Cohen, 1986).

The same technology has allowed the construction of so-called "massively parallel" computers. In a traditional computer, there is only one processor dedicated to instruction decoding and to the execution of arithmetic and logical operations. This central processing unit (CPU) interacts with a large memory, which acts as a passive recipient of data.

The paradoxical situation of a traditional computer is that only a small fraction of the silicon present is used in an intensive manner, while the remaining portion, more than 90%, is almost always inactive (Hillis, 1985). Besides, there is an obvious bottleneck in communications between CPU and memory, which creates problems of efficiency in data intensive applications.

Massively parallel computers (Baiardi and Vanneschi, 1987) substitute the architecture based on a single CPU plus a large memory with a "community" of interacting CPU's. According to the particular approach used, each CPU can have its own private memory, or all of them may access the same memory. The former solution is the most radical one, since it eliminates the bottleneck of communication with a single memory.

Massively parallel computers are particularily attractive since they can provide huge computing power at reasonable cost. However, several algorithms which were devised for sequential computers behave unsatisfactorily on highly parallel machines: this can be due either to the sequential nature of the operations involved, or to the overhead introduced by communications among the different processors.

The search for effective parallel algorithms represents a crucially important issue for the development of massively parallel systems. The dynamical systems described above are intrinsically parallel, and thus represent a promising answer to this requirement (Recce and Treleaven, 1988).

A further reason for interest in neural networks lies in the discovery of algorithms which enable learning in multi-level networks without it being necessary (at least in principle) to predefine the internal architecture, as in the case of perceptrons. These algorithms, which overcome some limitations of classical perceptrons, will be discussed in Chap. 5.

As we have seen, a part of this recent interest in neural networks comes from researchers in disciplines which are traditionally extraneous to this field. Also among researchers in artificial intelligence, however, there is a significant growth of interest, which is fairly surprising in the light of past experience. In fact, while researchers in dynamical systems and statistical mechanics consider these as new subjects, those working in AI could regard them as old ideas which have already been analyzed, and abandoned, in the development of their own field.

The reasons for this renewed popularity of neural networks in the AI community are, to some extent, those described above: greater computing power, possibility of producing "neural" chips, massively parallel machines, interest in nonlinear systems, the overcoming of some limits of the classical perceptron. But we believe that one of the fundamental reasons for the interest in the dynamical approach should also be found in the limitations of classical AI, which are now much better understood than in the past.

Classical AI has obtained important results in the computer simulation of complex cognitive processes such as, for example, medical diagnosis, which have in their turn led to great expectations both in the scientific and industrial fields. Once again, however, things turned out to be more difficult than the enthusiasts tended to think. Thus, there is currently a tendency to see dynamical systems, now more credible and interesting for the reasons cited above, as a possible solution to some of the difficulties in AI.

2.4 The Relationship with Classical AI

In this section we will begin to deal with the relationships between two different lines of research directed towards the realisation of artificial cognitive systems: classical AI and dynamical approach. This subject will be taken up again in Chap. 7, with the hindsight obtained from the analysis of various models, carried out in Chaps. 4, 5 and 6. In this section we will, above all, underline the problems of AI and the hopes raised by the dynamical systems approach, whereas the limits of the latter will be discussed in Chap. 7.

It is known that artificial intelligence has produced some important theoretical results, and several smart applications. For example, there are currently available "expert systems", computer programs capable of reaching the performance of a human expert, in "difficult" sectors such as medical diagnosis, chemical analysis, geological prospecting, process control, interpretation of legal texts, amongst others.

Although the various currents of thought present in AI give rise to a lively and intense debate, it is possible to outline some of the fundamental aspects of classical AI which the various schools hold in common (with the partial exception of those studying image processing).

The first common aspect is the choice of operating with "entities" of a high conceptual level. There is no attempt to model the operation of particular physical systems, but direct use of symbols is made to represent concepts.

This choice of operating with high level abstract entities is one of the main reasons for the success of AI: this does, in fact, allow one to directly attack problems considered to be difficult, to avoid getting bogged down in the attempt to represent "simple" concepts and relationships in a well-defined structure (a neural network, for example). The terms "difficult" and "simple" are used here in an intuitive sense, related to the perception of difficulties in day-to-day experience.

The basic characteristic of classical AI is the use, for the manipulation of these symbols, of chains of inferences or "rules" of the "if A, then B" type, as in logic programming or production systems. AI systems reach their goal through chains of inferences, that is through explicit reasoning.

The introduction of any particular rule has a "semantic" justification, that is it draws its motivation from the meaning of the symbols it manipulates (for example, "X father of Y", "Y parent of Z" $\Rightarrow$ "X grandfather of Z"). This entails a strong dependence of the rules upon the specific problem being dealt with.

AI systems also show a centralized "control", i.e. a set of rules (the so-called meta-rules) which are used to decide which rules of inference to apply in a given computational stage. The meta-rules are of a more general nature than the production rules, and depend only to a limited extent upon the domain of application.

Another aspect common to the various approaches to classical AI is the use of dedicated symbols to represent each concept (local representations). In this way, there must be a symbol for each concept.

On the contrary to AI programs, dynamical systems reach their conclusions by applying evolutive rules to numerical variables, instead of applying inference rules to logical variables. Also, in the systems considered these evolutive rules are the same for each element and are justified on the basis of plausibility considerations, neurophysiological analogies and, above all, a posteriori, because of their effectively demonstrated capacity to memorise and generalise. The dependence upon the specific case being examined, in these systems, is usually reflected in the choice of suitable values of the parameters, while the form of the dynamical equations remains constant.

Dynamical systems also lack a "control" in the classical sense, and all the units are simultaneously working (that is, they continuously compute their state, without being called at work by a central controller). While in classical systems there is a "conductor", the control, which establishes "who" is authorised to operate at any particular instant, in dynamical systems the order reached derives from direct interactions amongst the elementary units.

Moreover, while every rule in an AI system has a precise, pre-assigned meaning, it is fairly common that several neurons in a dynamical model have no pre-defined meaning: they specialize, during the learning phase, in a manner which is often unexpected and sometimes difficult to understand for the network designer.

While classical AI systems adopt local representations, an alternative, preferred by many researchers of the dynamical systems approach, consists of the so-called distributed representations, where high level concepts can be defined as activation patterns over several nodes.

Using distributed representations it is not necessary to identify a priori every meaningful variable: even without any previous knowledge, the study of a certain number of specimens of a category can lead to the formation of co-activation patterns corresponding to that concept, given that the learning algorithm is adequate.

It should be noted, however, that it is possible to obtain dynamical models for AI also by utilising a "formal neuron" for each concept (the so-called "grandmother neuron", which designates, in the jargon of neurophysiologists, the hypothesis that there is a specific neuron for every concept).

Finally, it should be noted that while AI systems possess a well-defined termination condition, the goal, for dynamical systems this is unnecessary. It is certainly possible to introduce conditions of this kind even in dynamical systems, by labelling, for example, the terminal states with suitable variables, but it is also possible to let the system evolve towards the asymptotic state, and consider the latter as a terminal state.

In conclusion, it can be stated that dynamical systems perform cognitive tasks by relying on their self-organizing properties much more than classical AI systems do. These latter, in their turn, rely on spontaneous organizational properties much more than traditional software engineering techniques do.

In fact, software science has evolved by following a classical engineering paradigm, emphasizing the decomposition of complex program into modules, which are carefully engineered in order to perform their function and to fit perfectly whith the other modules. Moreover, the programmer can also predict the order in which different parts of the program will be executed. Unpredictability of the artificial system, in this context, is undesired and dangerous, and manifests itself only through errors, or "bugs".

Artificial intelligence emphasizes the separation of declarative knowledge (i.e., the rules which describe the domain) from the actual working of the program. Ideally, the programmer should focus on providing the correct rules, leaving the choice of the order of their application to an automatic procedure (the control). In this way, the programmer gives up the idea of knowing exactly what the program is doing at every time.

However, in AI systems every rule has a well defined meaning, and the programmer exactly knows why that rule was introduced. As we have seen, in the dynamical approach much more room is left for unpredictability: a given node can have no predefined meaning, and its role emerges through self-organization.

Besides the notable successes obtained, AI must face several very serious problems, important both from the theoretical and applications viewpoints.

One of the main reasons for the interest in cognitive dynamical systems arises from the consideration that they show promising behaviour in precisely those aspects which are the most problematic ones for AI.

Let us now briefly examine some of the principal problems of classical AI. For definiteness, we will refer mainly to a particular sector of AI, the one which has provided the greatest successes over the past few years, that is, knowledge based systems (or expert systems). The major problems, which will be discussed below, concern:

- limited domain
- knowledge engineering
- contradictions and time varying knowledge
- learning
- brittleness.

Experience has shown how it is possible to obtain sometimes good, and even spectacular, results in limited fields of knowledge. Consider, for example, the numerous expert systems which have been successfully developed in the fields of computer configuration, medical diagnosis, chemistry, process control.

Previously, AI research was focussed upon systems of a general nature, "general problem solvers". Thus attention was paid especially to the study of general search methods, in the hope that these would avoid the "combinatorial explosion" in the various specific sectors. The experience accumulated in the meantime has led to a significant change of approach: it is now considered necessary to rely on a significant amount of specific information about the domain. The combinatorial explosion is avoided by drastically reducing the number of alternatives at each stage, through the introduction of a large quantity of problem-dependent information.

The need to specify the rules in extreme detail leads to a further increase in their number. Although hierarchical methods for knowledge representation do exist, which alleviate the computational burden, it has so far proved impossible to deal with very large knowledge bases. The limited domain of application of existing systems is one of the main problems of AI, both from the theoretical and applicative points of view.

The fundamental problem, deriving from this observation, is how it is possible to manage a very large amount of disparate information, in order to allow a rapid recall of only that which is pertinent to a given context. In other words, the problem is that of rapidly identifying the appropriate context. This is clearly exemplified in the understanding of natural language, where human beings succeed in mastering an enormous quantity of information, which can be recombined in various manners, with a rapidity and subtlety which is completely unknown to artificial systems.

Another critical aspect of AI systems is the elicitation of this specific knowledge about the problem under examination, which is crucial for the success of the system. Definition of the rules to be loaded into an expert system is a result of the process known as "knowledge engineering". As is well known, this is a procedure of successive refinements of the knowledge base, through the interaction of an expert in the domain of application and a "knowledge engineer", whose task is to find the most suitable representation formalism and to define the data structure and the correct rules to be loaded.

Knowledge engineering is thus a crucial task for the success of the project, and is often difficult and unnatural because it requires, above all, the making explicit of the knowledge and working methods of the expert, which are often implicit. The making explicit of implicit knowledge may have an important role in the development of methods for the sharing and formalisation of knowledge, but it is also rather unnatural and error prone. It is also a process which employs high-level specialists for long periods and is therefore quite expensive.

The difficulties and the costs of knowledge engineering are thus an important problem and an obstacle to the widespread diffusion of AI techniques. The importance of automated learning techniques thus becomes clear, and particularly of learning by examples. According to this method, the system is presented with a series of cases, together with the correct answers. Also within "classical" AI, research in automated learning is experiencing a period of great interest. Without intending to review the field in the present work (see e.g. Michalski et al., 1984 and 1986), we can note that most learning techniques based upon logical formalisms require a very accurate choice of examples and, in particular, that contradictory examples are excluded.

The problem of contradictions is a general one for all systems based upon logical formalism and not only for automated learning systems. If there are contradictions in the rules, the behaviour of the system becomes extremely unreliable.

Another well-known problem of AI systems concerns the handling of knowledge which varies with time. In this case, one deals, for example, with a statement which is true until a certain point in time, after which it ceases to be so. From that moment, therefore, there is no longer a guarantee of the validity of the inferences which had been demonstrated by using that statement. However, the retracing of all of these chains of dependencies is an extremely costly calculation. This problem of variable knowledge is closely related to that of contradictions: in fact, a variable knowledge base, if it does not correctly retrace all of the consequences of variations in truth values, may then contain contradictory statements.

The problem of the "brittleness" of knowledge based systems (Holland, 1986) should also be mentioned. This term refers to the fact that the quality of performance rapidly declines as soon as one leaves the specific field for which the system was designed. This is related to the need to provide a lot of problem specific knowledge in order to achieve a satisfactory performance.

After having outlined the main problems in classical AI, we will now show that dynamical cognitive systems are capable of providing some interesting stimuli for overcoming at least some of them.

First of all, dynamical systems possess mechanisms for dealing with contradictory knowledge. In fact, the evolutive law implicitly defines a rule for conflict resolution.

Let us take, e.g., the case of a formal boolean neuron, which receives inputs from other neurons, each of which is multiplied by the corresponding coupling coefficient, and compares the sum of these inputs with a threshold θ. The neuron fires (i.e., it takes the value "1") if the net input exceeds the threshold, while it remains quiescent otherwise.

If we concentrate on the i-th neuron, and if W_{ik} represents the intensity of the synapse which goes from the k-th to the i-th neuron, we have:

$$X_i(t+1) = H(G_i(X(t)) - \theta) \tag{2.5}$$

where

$$G_i(X) \equiv \sum_{k=1}^{N} W_{ik}X_k \ . \tag{2.6}$$

N is the total number of neurons in the network and $H(Y)$ is the Heaviside function which is 1 if $Y > 0$ and otherwise zero.

Let us consider a simple case, where the state of the i-th element corresponds to a precise classification of the input state: for example, $X_i = 1$ could mean that the input belongs to class A, whereas $X_i = 0$ indicates that it does not.

The i-th element will receive information from all the elements to which it is connected, and some of this information could be contradictory, pointing towards different classifications. The evolution rule automatically takes on the task of weighting the various pieces of evidence, which contribute towards the determination of the input functions, thus arriving at the conclusion.

It should be noted that the previous statement does not necessarily mean that the classification is a correct one: this depends upon the nature of the problem, the neural network model used and the training. In any case, however, the system is not impeded by contradictions. A system based upon explicit rules, in order to reach the same result, would have to define a voting mechanism for the different pieces of evidence or it would have to explicitly consider all possible combinations of the evidence, associating them with the corresponding classification.

Another extremely important aspect is that, as we have already seen, algorithms for automated learning by examples exist. This can eliminate, or drastically reduce, the bottleneck represented by knowledge engineering.

It should be added that learning-by-example can take place even in the presence of noise and contradictions within the training set. It may happen, for example, that the set of examples contains several analogous situations which have been classified in different ways, either by error or because they have been assessed by different teachers or at different times.

The learning mechanisms, which are also dynamical systems, automatically weight the various examples, and can thus operate efficiently even in the presence of such contradictions. Similar observations can be made for the problem of time-variable knowledge: while an expert system may find itself faced with an inconsistency, a dynamical system having reached a certain state will be simply shifted to a slightly different state. The evolution rule will thus determine the future history of the system. The management of variable knowledge and situations therefore comes naturally in dynamical systems.

The observations made so far imply a certain robustness of dynamical systems with respect to the presence of contradictions and noise, both in the data and in the "rules". Dynamical networks also exhibit fault tolerance: if an element does not function properly – either due to actual faults in the hardware, if the network is directly implemented in hardware, or to programming errors, if it is simulated on a digital computer – this may influence the system, but very often this influence is quite small. This is a consequence of the fact that one is computing with dynamical attractors (Hogg and Huberman, 1984), that is with global properties of the system.

As will be shown in the following chapters, analogous considerations can be made about the processing of incomplete input data: it is sufficient that the input resembles the input of a case learnt during training, to ensure that the system evolves towards the corresponding attractor.

It can also be seen that dynamical systems possess the ability to generalise. The choice of a certain set of variables determines the phase space of the system. The learning phase then determines the structure of the phase space, that is the set of attractors along with their basins of attraction, and every initial state is associated with the corresponding asymptotic one. With a limited number of examples it is then possible to cover the whole range of possible cases. Dynamical systems can therefore "propose hypotheses" also in situations quite remote from those met during training.

To summarise, the characteristics of robustness and of resistance to noise, contradictions and incompleteness, together with the possibility of automated learning, are the most attractive ones of dynamical systems compared to inferential systems.

One point should be noted: so far, the robust nature of dynamical systems has been emphasized. This property should not be taken in an absolute sense, as it is known that there are chaotic systems whose evolution shows a marked dependence upon the initial conditions.

It is not yet clear, however, whether it will be possible to deal with very wide knowledge domains using cognitive dynamical systems. So far, to our knowledge, there are no examples of this kind. However, at least one reason for these domain limitations derives from the essentially sequential nature of classical systems and, undoubtedly, dynamical systems are intrinsically parallel.

Of course, the dynamical approach to AI is not free from drawbacks. Some tasks, which are easily carried out using the symbolic approach, are extremely difficult using dynamical systems. A detailed discussion of the limits of this approach is to be found in Chap. 7, following the analysis of the various specific models.

3. Dynamical Behaviour of Complex Systems

3.1 Introduction

This chapter introduces some concepts of a general nature about the behaviour of dynamical systems, with particular attention being paid to those classes of systems whose cognitive properties will be studied in the following chapters. Artificial cognitive systems often are built on reasonably simple elementary structures, so that it is possible and worthwhile developing a quantitative treatment. On the other hand, as we have already mentioned and as we shall see later, even simple elements can lead to complex and unforeseeable behaviours.

A particular class of dynamical systems will be studied: discrete-time systems, i.e. systems described by difference equations instead of differential equations. This choice is linked to the fact that the states of dynamical systems to be dealt with in the following chapters are updated in a discrete manner, that is, in correspondence with discrete values of time (a very natural choice when dealing with digital computers).

Systems of the following type (iteratively defined) will be studied:

$$x(t+1) = f[x(t), P, t] \quad . \tag{3.1}$$

In the system described by Eq. (3.1), the vectors x, f and P are defined as

$$x = \{x_1, x_2, \ldots, x_N\}$$
$$f = \{f_1, f_2, \ldots, f_N\}$$
$$P = \{P_1, P_2, \ldots, P_M\} \quad .$$

This system describes the dynamics of the vector x starting from the initial condition

$$x(0) = x_0 = \{x_{01}, x_{02}, \ldots, x_{0N}\} \quad .$$

In general, the second member f will be nonlinear in the components x_j of the vector x.

The system (3.1) takes into account the possibility of a coupling with the external environment not only through the initial conditions, but also through an explicit time dependence of the right-hand side: that is, through a "forcing function". When this explicit time dependence is absent, the system (3.1) is called "autonomous", while the term "non-autonomous" is used for the opposite case.

In principle, also the vector of the parameters P may be time-dependent. This is the case, for example, of a dynamical network in which the connections between the elements are modified during time evolution. Even for P, however, it is necessary to distinguish between an explicit time dependence and an implicit one through the vector x, as, for instance, in a network in which the connections between the elements are modified according to the values taken by a suitable function of the state variables.

The study undertaken in this chapter will concern a potentially infinite number of successive time steps, in order to show the full range of asymptotic behaviours typical of dynamical time-discrete systems. Not all artificial cognitive systems, however, rely upon asymptotic dynamics: in particular, in Chap. 5 some networks are discussed whose cognitive behaviour is not associated with a truly asymptotic dynamics.

This chapter will study first of all the case of discrete dynamical systems with only one state variable (Sect. 3.2). With reference to such systems, the study will include the relaxation towards fixed points, the appearance of periodic attractors ("limit cycles") and the development of "chaotic" situations. The ideas will be brought into focus by referring to particular equations which, however, show a sufficient range and variety of behaviours: the "triangular map" and the "logistic map". The general concepts of attractor, deterministic chaos, Lyapunov exponent and fractal dimension will be introduced with reference to these systems.

The extension to multiple degrees of freedom will be studied in Sect. 3.3 with special reference being made to one particular system: the so-called "Hénon map", which consists of a generalisation of the "logistic map". The concepts of dissipative systems and strange attractors will be introduced in relation to the Hénon map.

The study of discrete dynamical systems with multiple degrees of freedom will then be further extended by referring to a particular class: that of systems which are also discrete in the value of their state variables; more precisely, boolean systems, in which the state variables can only take the values 0 or 1. This choice will also provide a direct link with the following chapter, in which the cognitive properties of particular boolean dynamical networks will be studied.

Boolean dynamical systems with multiple degrees of freedom may be divided into various classes as a function both of their degree of connectivity (i.e. of the number of state variables at time t upon which each one depends at time $t + 1$) and of the topological characteristics of the connectivity itself. In fact, as we shall see later, some types of connections allow a natural representation of the state variables on a line, whereas other types of connections require a two-dimensional representation. Complete connectivity (which, as we shall see in Chap. 4, is a characteristic of Hopfield networks) constitutes a particular case, since it is associated with no particular topology.

We shall limit ourselves, in the present chapter, to the study of low connectivity systems, in which the variable associated to each degree of freedom changes its state, at time $t + 1$, according to a boolean function of the state of its nearest neighbours at time t. Even in relatively simple cases, however, it is not possible to develop an analytical theory of these systems: the results to be discussed are therefore mostly obtained from computer experiments.

The most simple case of low connectivity is the one-dimensional one, where each degree of freedom is only influenced by its nearest neighbours. As examples of this case we shall study cellular automata (Sect. 3.4). In these structures, we shall again see a great variety of dynamical behaviours: steady states (homogeneous and non-homogeneous), oscillatory behaviours, aperiodic behaviours giving rise to fractal structures.

As an example of a two-dimensional system with limited connectivity, we shall consider the so-called "life game" (Sect. 3.5). This is a system defined by quite a simple rule for the dependence of every state variable upon its eight nearest neighbours, a rule of biological inspiration. Notwithstanding its simplicity, this microscopic rule gives rise to an unforeseeable variety of dynamical macroscopic behaviours. In this case, we meet not only stationary or pulsating structures, but also propagating ones ("gliders");combinations of all the previous structures are also observed (for example, complex structures which evolve in different ways, varying their spatial extension without any apparent periodicity, and "shooting out" gliders). The interactions, destructive or constructive, between different objects then give rise to quite a varied landscape. It has been widely studied at a phenomenological level without, however, succeeding to develop a unitary theoretical framework.

One may speak of "life game" not only because the choice of the microscopic rule which governs its functioning is inspired by a biological metaphor, but also because, in the two-dimensional array which constitutes the "universe" of this game, not only single cells but also "organisms" are born, grow and die. The latter show unpredictable properties which can not be described at the elementary level which does, however, support them.

In Sect. 3.6, random boolean networks will then be introduced. This is a particular class of two-dimensional non-homogeneous cellular automata with interesting self-organizing properties.

All the networks studied so far have the property of being made up of extremely simple cells, each characterised by a single state variable. The reaction-diffusion systems which will be briefly introduced in Sect. 3.7 are, on the other hand, examples of the cognitive capabilities of a dynamical network having a well-defined topology in which each element is defined by a multivalued, multi-component state vector.

3.2 One-Dimensional Dynamical Systems

In this section we shall consider autonomous dynamical systems of the type

$$x(t+1) = f[x(t), P] \tag{3.2}$$

which from now on will be written

$$x^{(i+1)} = f[x^{(i)}, P] . \tag{3.3}$$

It will be said that a system of the type (3.2) has a fixed point x^* if the relationship

$$x^* = f(x^*) \tag{3.4}$$

holds.

In geometrical terms, the fixed points are thus found at the intersection between the function f and the bisectrix of the angle between the two Cartesian axes which respectively represent $x^{(i)}$ and $x^{(i+1)}$.

From the time-evolution point of view, there is a fixed point x^* if, given $x^{(0)} = x^*$, one has $x^{(i)} = x^*$ for every value of i. A fixed point is said to be stable (more precisely, asymptotically stable) if, for all the initial conditions belonging to a suitable neighbourhood of x^*, one has

$$\lim_{i \to \infty} x^{(i)} = x^* \tag{3.5}$$

or, in other terms, if

$$\lim_{i \to \infty} f(f(\ldots f(x^{(0)})\ldots)) = x^* . \tag{3.6}$$

From this definition, one immediately obtains a criterion for asymptotic stability. Let us consider, in fact, the distance between the $(i + 1)$-th iteration $x^{(i+1)}$ and the fixed point x^*

$$\delta^{(i+1)} = |x^{(i+1)} - x^*| . \tag{3.7}$$

This expression may be written as:

$$\delta^{(i+1)} = |f(x^{(i)}) - x^*| = \left| \frac{f[x^* + (x^{(i)} - x^*)] - f(x^*)}{x^{(i)} - x^*} \right| \delta^{(i)} \tag{3.8}$$

and, if the function f can be differentiated

$$\delta^{(i+1)} \approx |f'(x^*)| \delta^{(i)} . \tag{3.9}$$

But, since asymptotic stability requires that

$$\lim_{i \to \infty} \delta^{(i)} = 0$$

this implies

$$|f'(x^*)| < 1 \quad .\tag{3.10}$$

Let us now consider, as an example of a one-dimensional autonomous dynamical system which depends upon a single parameter, the so-called "triangular map" (Schuster, 1984), defined as

$$x^{(i+1)} = A\left\{1 - 2|\tfrac{1}{2} - x^{(i)}|\right\}\tag{3.11}$$

with $0 \le A \le 1$ and $0 \le x^{(0)} \le 1$. Fig. 3.1 shows a geometrical representation of the map.

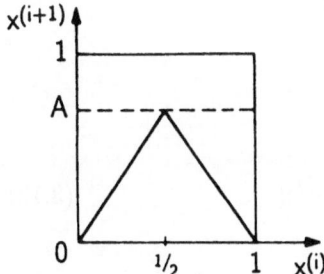

Fig. 3.1. The triangular map. (After Schuster, 1984)

The stability criterion introduced above allows us to state that, for $A < 1/2$, the origin ($x^* = 0$), which is the only fixed point of the map, is asymptotically stable. This may be immediately verified, as Fig. 3.2a shows, using a graphical representation of the results of successive iterations starting from an initial condition $x^{(0)} \ne 0$. In this interval of values for A, the origin is asymptotically stable for all possible initial conditions with $0 \le x^{(0)} \le 1$.

When $A > 1/2$, however, there are two unstable fixed points, one of which being the origin. This can be verified graphically. Fig. 3.2b shows the

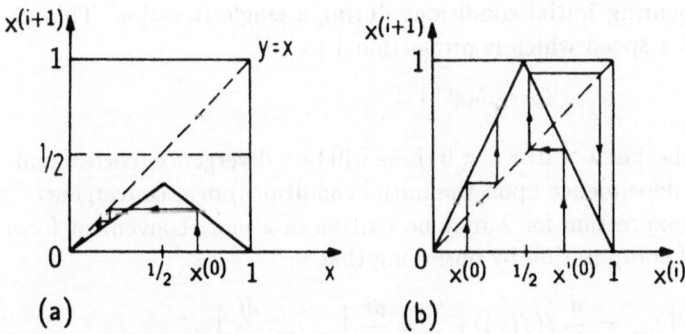

Fig. 3.2. a) $A < 1/2$: one stable fixed point $x^* = 0$; b) $A = 1$: two unstable fixed points. (After Schuster, 1984)

case where $A = 1$, with the corresponding unstable fixed points at 0 and 2/3. The instability is clear from the behaviour of the successive iterations corresponding to the initial conditions $x^{(0)}$ and $x'^{(0)}$ shown in the figure.

In the absence of stable fixed points, the results of iterations corresponding to closely neighbouring initial conditions may progressively move away from one another. For an analytical characterisation of this situation, it is useful to introduce the concept of Lyapunov exponent (Schuster, 1984; Bergé et al., 1984).

Let us consider two close initial conditions

$$x^{(0)}, \quad x^{(0)} + \varepsilon \quad .$$

After n iterations of the map f, the values obtained are

$$f^n(x^{(0)}), \quad f^n(x^{(0)} + \varepsilon) \quad .$$

The Lyapunov exponent

$$\lambda(x^{(0)})$$

is defined through the relationship

$$\varepsilon e^{n\lambda} = |f^n(x^{(0)} + \varepsilon) - f^n(x^{(0)})| \quad . \tag{3.12}$$

If the function f is a regular one, in the limit

$$\varepsilon \to 0, \quad n \to \infty$$

one obtains

$$\lambda(x^{(0)}) = \lim_{n \to \infty} \lim_{\varepsilon \to 0} \frac{1}{n} \ln \left| \frac{f^n(x^{(0)} + \varepsilon) - f^n(x^{(0)})}{\varepsilon} \right|$$

$$= \lim_{n \to \infty} \frac{1}{n} \ln \left| \frac{df^n(x^{(0)})}{dx^{(0)}} \right| \quad . \tag{3.13}$$

The Lyapunov exponent λ thus provides a measure of the divergence between two neighbouring initial conditions during a single iteration. This divergence grows at a speed which is proportional to

$$e^{\lambda(x^{(0)})} \quad .$$

According to whether $\lambda > 0$ or $\lambda < 0$ there will be a divergence (corresponding to a sensitive dependence upon the initial conditions) or a convergence.

The preceding expression for λ may be written in a more convenient form for the purpose of computation by observing that

$$\frac{d}{dx} f^2(x) \Big|_{x^{(0)}} = \frac{d}{dx} f(f(x)) \Big|_{x^{(0)}} = \frac{df}{dx} \Big|_{f(x^{(0)})} \frac{df}{dx} \Big|_{x^{(0)}}$$

$$= f'(x^{(1)}) \cdot f'(x^{(0)}) \quad . \tag{3.14}$$

This relationship can immediately be generalised to the case of n iterations, so that

$$\lambda(x^{(0)}) = \lim_{n \to \infty} \ln \left| \frac{d}{dx^{(0)}} f^n(x^{(0)}) \right| = \lim_{n \to \infty} \frac{1}{n} \ln \left| \prod_{i=0}^{n-1} f'(x^{(i)}) \right|$$

$$= \lim_{n \to \infty} \frac{1}{n} \sum_{i=0}^{n-1} \ln |f'(x^{(i)})| \quad . \tag{3.15}$$

By applying this definition to the triangular map, one obtains an expression independent of $x^{(0)}$ for the Lyapunov exponent:

$$\lambda = \ln 2A \quad .$$

Thus,

$$\lambda < 0 \quad \text{for} \quad A < 1/2$$

(this condition, as we already know, corresponding to the case of one stable fixed point) and

$$\lambda > 0 \quad \text{for} \quad A > 1/2 \quad .$$

The study through Lyapunov exponents of the behaviour of solutions therefore confirms that the map (3.11) undergoes a transition at $A = 1/2$. From a dynamics of asymptotic convergence towards the origin it changes to an unpredictable behaviour, in the sense that the proximity of the initial conditions does not lead at all to a nearness during the successive evolution.

In this case, then, there is a "sensitive dependence" upon the initial conditions, and "deterministic chaos" arises (Schuster, 1984). This chaotic behaviour, however, is not due to the stochastic nature of the equation or to the presence of external noise; it is rather due to the exponential separation between trajectories and to the simultaneous limitation of the range of possible values for the state variables (which, in this specific case, must lie between zero and one).

Knowledge of the initial conditions could then allow prediction of the successive behaviour only if the former were known with infinite precision. The unpredictability of deterministic systems with chaotic behaviour then arises from the fact that any real observation is characterised by a finite precision.

The triangular map has allowed the introduction of some general concepts and has showed the possible rise of a situation of deterministic chaos. This map, however, has a rather limited range of possible behaviours: in particular, it does not show periodic solutions. For this reason, we shall examine another system with a single degree of freedom but with a greater variety of behaviours: the "logistic map".

The logistic map (Schuster, 1984) is a dynamical system with one degree of freedom depending upon a single parameter:

$$x^{(i+1)} = Ax^{(i)}(1 - x^{(i)}) \tag{3.16}$$

with $0 \leq x^{(i)} \leq 1$ for each i.

The map (3.16) is often called "logistic map" with reference to a possible ecological interpretation. If, in fact, x is the number of individuals of a certain species, it may be reasonably stated that, at time $(t + 1)$, x is determined by the difference between a linear term in $x(t)$ ("generation" term) and a quadratic term in $x(t)$ ("competition for vital resources" term).

Let us consider the fixed points of Eq. (3.16) and study their stability. First of all, it can be seen that the condition $0 \leq x^{(i)} \leq 1$ for each i bounds the possible values for A between $0 < A \leq 4$. In fact, for $A < 0$, positive values of $x(i)$ would correspond to negative values of $x(i + 1)$; for $A > 4$, we should have $x(i + 1) > 1$ over the whole range of values for $x(i)$.

As shown in Fig. 3.3a, for $0 < A < 1$ the parabola corresponding to the rhs of (3.16) intersects the bisectrix of the first quadrant only at the origin. Moreover, recalling the stability condition (Eq. (3.10)), we have

$$\left| \frac{df(x)}{dx} \right|_{x=0} = A \tag{3.17}$$

and therefore, for $0 < A < 1$, the origin is also asymptotically stable.

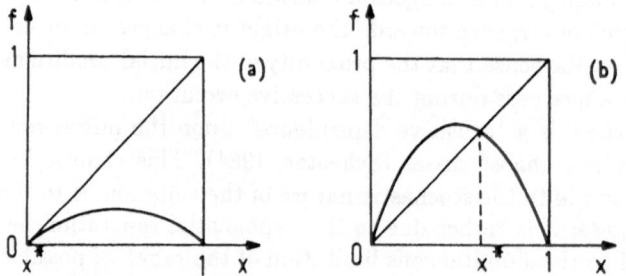

Fig. 3.3a,b. The fixed points of (3.16) for a) $A < 1$; b) $1 < A < 3$. (After Schuster, 1984)

When $A > 1$, besides the origin there is also the fixed point $x^* = 1 - 1/A$ (see Fig. 3.3b), for which

$$\left| \frac{df(x)}{dx} \right|_{x=1-1/A} = |2 - A| \quad . \tag{3.18}$$

This expression is less than unity for $1 < A < 3$, and therefore the second fixed point is stable. The quantity (3.18) becomes greater than one for $3 < A < 4$; over the latter range of values of the parameter A, therefore, also the second fixed point becomes unstable.

To increase our understanding of the behaviour of the map in the range $3 < A < 4$, let us consider the second iterate $f^2(x)$. This is a fourth degree curve, of the type shown in Fig. 3.4.

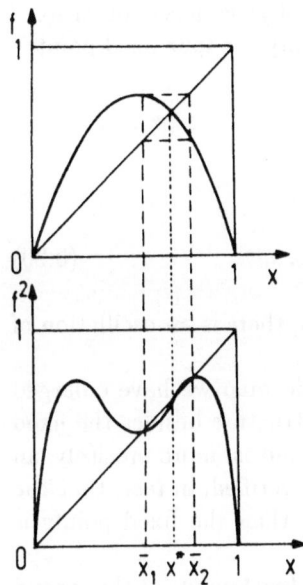

Fig. 3.4. The map $f(x)$ and the iterate $f^2(x)$ for $A > 3$. (After Schuster, 1984)

First of all, it can be shown that the origin and $x^* = 1 - 1/A$ are fixed points of f^2. A fixed point for f does, in fact, remain so for all the iterates. Let us now study the derivative of f^2 at those fixed points, observing that

$$\frac{df^2}{dx^*} = f^{2\prime} = [f'(x^*)]^2 \ . \tag{3.19}$$

For $A > 1$, the origin is unstable. For $A > 3$, the point $x^* = 1 - 1/A$ also becomes unstable, because $f^{2\prime}(x^*) > 1$. But, as Fig. 3.4 shows, there are two new points of intersection

$$\bar{x}_1, \ \bar{x}_2$$

between the second iterate and the bisectrix. At these points, for geometrical reasons, $f^{2\prime}$ must be smaller than 1. These are, therefore, fixed points of the second iterate.

Let us now examine the relevance of these stable fixed points of the second iterate from the point of view of the logistic map (Eq. (3.16)). It can be observed that the relationship

$$f^2(\bar{x}_1) = \bar{x}_1$$

implies

$$f f[f(\bar{x}_1)] = f[f^2(\bar{x}_1)] = f(\bar{x}_1)$$

and the same can be said for $\bar{x}_2$.

The iterate through f of one of these fixed points of f^2 is thus another fixed point of f^2. This map, as we have seen, possesses only the four fixed points

$$0, \ x^*, \ \bar{x}_1, \ \bar{x}_2$$

and it necessarily follows that

$$f(\bar{x}_1) = \bar{x}_2, \quad f(\bar{x}_2) = \bar{x}_1 \ . \tag{3.20}$$

In other words, starting from any of these points, there is an oscillation of period two between them.

It can thus be seen that by studying the logistic map we have enlarged the field of possible attractors: there is a periodic structure besides the fixed points. Also this periodic structure is a stable one, and is, more precisely, an asymptotically stable "limit cycle". It can be easily verified, in fact, that the Lyapunov exponent for any initial condition other than the fixed points is negative.

On further increasing the value of A, also the new fixed points of the second iterate

$$\bar{x}_1, \ \bar{x}_2$$

become unstable. It is then necessary to consider the fixed points of the fourth iterate and their stability. As can be seen from Fig. 3.5, besides the preceding fixed points, which have now all become unstable, there are four more. In a similar way to that seen for the second iterate, these four new fixed points give rise to an asymptotically stable limit cycle of period four.

This behaviour repeats itself in a regular way and, on further increasing the value of the parameter A, the period of the limit cycle continues to double. The results of a numerical simulation are given in Fig. 3.6 (Crutchfield et al., 1982). Figure 3.6a shows the points which make up the various limit cycles as a function of the parameter A. It can be seen that the values A_n, which correspond to the transitions towards cycles of doubled periods, get closer and closer to each other, and have an accumulation point corresponding to the value

$$A_\infty = 3.5699 \ldots$$

which can be determined numerically.

The crowding of the transition points A_n follows the rule (which can be obtained theoretically)

$$\lim_{n \to \infty} \frac{A_n - A_{n-1}}{A_{n+1} - A_n} = \delta, \quad \delta = 4.6692 \ldots \ . \tag{3.21}$$

Figure 3.6b shows the Lyapunov exponent as a function of A. It can be seen that its value is always < 0 independently of the initial conditions ($= 0$

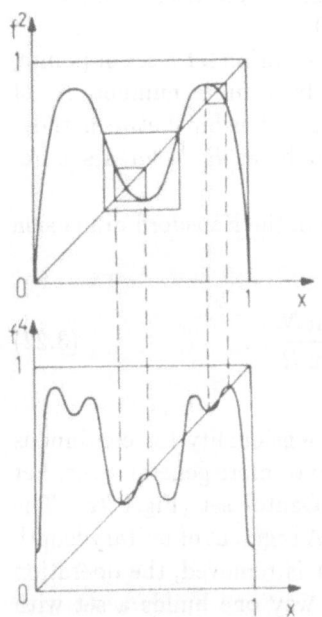

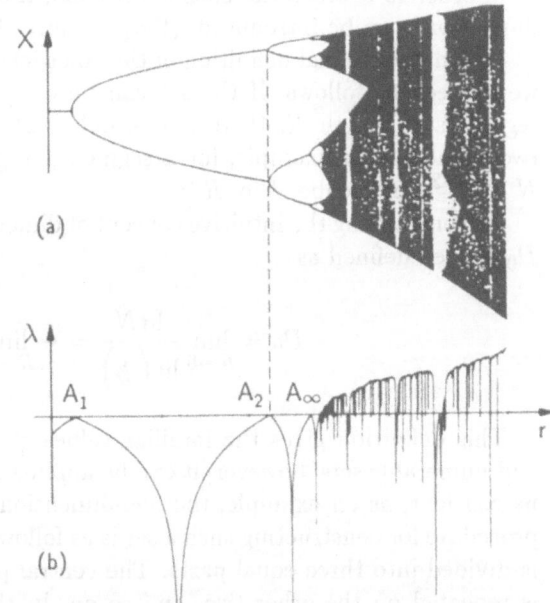

Fig. 3.5. The iterate $f^4(x)$ and the birth of a cycle of period 4. (After Schuster, 1984)

Fig. 3.6. a) Coordinates of points in the limit cycles of the iterates as functions of A; b) the Lyapunov exponent λ as a function of A. Note the change of sign of λ when $A = A_\infty$, corresponding to the transition to chaos (after Crutchfield et al., 1982)

at the transition points) for

$$A < A_\infty$$

while it turns positive immediately beyond that value. As we have already learnt from the study of the triangular map, this corresponds to the birth of a chaotic situation.

However, on examining more closely the Lyapunov exponent as a function of A when $A > A_\infty$, we see "windows" of negative values for the exponent. Numerical simulation shows that these windows correspond to periodic behaviours with periods which differ from the previous ones and successively double (or triple, quadruple, etc.).

It is now worthwhile introducing some considerations about the dimension of the attractors. The fixed points are obviously a zero-measure set on the x axis, as are the points corresponding to any periodic attractor. When, however,

$$A \to A_\infty$$

one has an attractor consisting of an infinite number of points. It is not, therefore, trivial to assign a dimension zero or one to such an attractor.

In order to characterise such a situation, the concept of "Hausdorff fractal dimension" can be introduced. (Bergé et al., 1984).

For an operational definition of the Hausdorff dimension for a set of points, we proceed as follows. If the set can be covered by a finite number, N, of segments of length R, there is a simple scaling relationship between these two numbers: for example, for a segment, N growths as R^{-1}; for a square, $N \propto R^{-2}$, for a cube, $N \propto R^{-3}$.

By generalizing the intuitive concept of dimension, the Hausdorff dimension D_0 is then defined as

$$D_0 = \lim_{R \to 0} \frac{\ln N}{\ln \left(\frac{1}{R} \right)} = - \lim_{R \to 0} \frac{\ln N}{\ln R} \quad . \tag{3.22}$$

This definition gives the familiar values of dimensionality for continuous and numerable sets; however, it can be applied also to more general cases. Let us consider, as an example, the one-dimensional Cantor set (Fig. 3.7a). The procedure for constructing such a set is as follows. A segment of unitary length is divided into three equal parts. The central part is removed, the operation is repeated on the other two, and so on. In this way one builds a set with a self-similar structure, i.e. with a structure which replicates itself in all its substructures.

For a self-similar set such as the Cantor one, the value of D_0 can be directly found from the construction algorithm. In fact, one can verify that $R = (1/3)^m$, $N = 2^m$. The Hausdorff fractal dimension for the Cantor set is thus

$$D_0 = \lim_{m \to \infty} \frac{\ln 2^m}{\ln 3^m} = \frac{\ln 2}{\ln 3} = 0.6309 \ldots \quad .$$

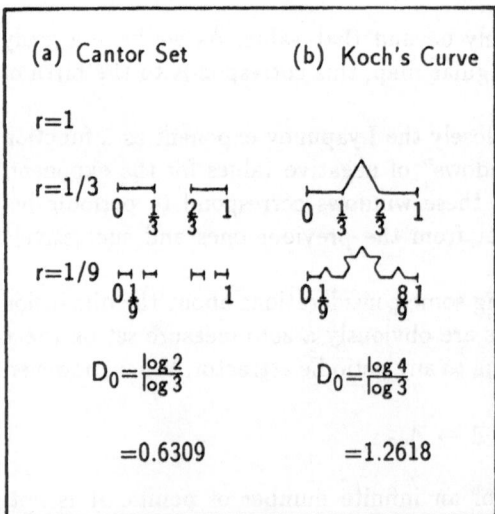

Fig. 3.7. a) First steps in the construction of Cantor set; b) first steps in the construction of Koch curve

As a further example, let us take the Koch curve (Fig. 3.7b), which is obtained by starting from a segment of unitary length, dividing it into three parts, substituting the central portion with the other two sides of an equilateral triangle whose side is 1/3 unit long, and so on. This gives another self-similar structure whose fractal dimension (Bergé et al., 1984) is

$$D_0 = \frac{\ln 4}{\ln 3} = 1.2618 \quad .$$

It is not surprising that, in this case, the fractal dimension lies between 1 and 2. Take, for example, the "snow flake" constructed by starting from an equilateral triangle with a side of unit length by using the Koch procedure (Fig. 3.8). As the number of steps $N \to \infty$, one obtains a curve of infinite length which encloses a finite area: this property is, in fact, expressed by a fractal dimension between 1 and 2.

Let us now go back to the structure of the attractors on the x axis for the logistic map as a function of the parameter A (Fig. 3.9). It can be seen that for $A < A_\infty$ one has zero-measure attractors. In the case where $A \to A_\infty$, a numerical evaluation of the Hausdorff dimension of the attractor gives the value $D_0 = 0.5388$ (Schuster, 1984).

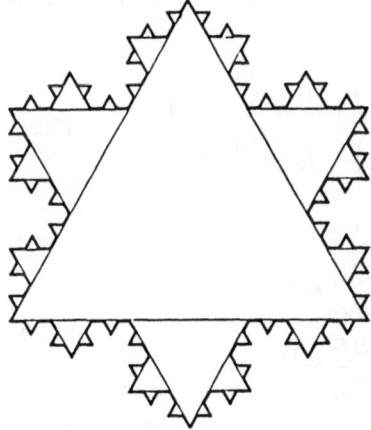

Fig. 3.8. A "snow flake" obtained after some steps in the construction of Koch curve

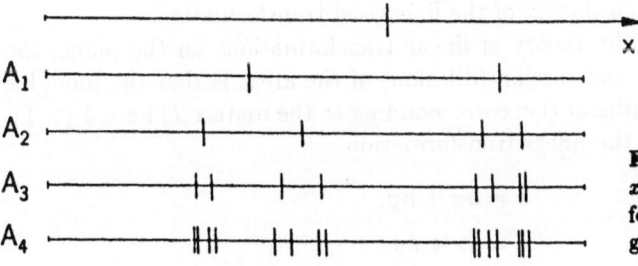

Fig. 3.9. Positions on the x-axis of the cycle elements for the iterates of the logistic map

3.3 Two-Dimensional Dynamical Systems

Consider a general two-dimensional autonomous system

$$x^{(i+1)} = f(x^{(i)}, P)$$
$$x = \{x_1, x_2\}, \quad f = \{f_1, f_2\} \ . \tag{3.23}$$

We have seen that, for one-dimensional systems, the contraction of the segments provides an indication on the stability of the attractors and, at the same time, on the type of dependence upon the initial conditions. We will now attempt to extend this result to the two-dimensional case in the simplest possible manner. Later on, however, we shall see that this extension is only a partial one which needs to be further studied.

We will now determine the conditions for a contraction of the areas in the plane. A two-dimensional system is called dissipative (with reference to the physical meaning of the word) if through the iterations there is a contraction of the areas in the plane. To obtain a mathematical condition of dissipativity, let us consider a small increment

$$\delta x^{(i)} = \left\{ \delta x_1^{(i)}, \delta x_2^{(i)} \right\} \ .$$

One then has (without explicitly mentioning the dependence upon P, for reasons of simplicity)

$$\delta x_1^{(i+1)} = f_1\left(x_1^{(i)} + \delta x_1^{(i)}, x_2^{(i)} + \delta x_2^{(i)}\right) - f_1\left(x_1^{(i)}, x_2^{(i)}\right)$$
$$\delta x_2^{(i+1)} = f_2\left(x_1^{(i)} + \delta x_1^{(i)}, x_2^{(i)} + \delta x_2^{(i)}\right) - f_2\left(x_1^{(i)}, x_2^{(i)}\right) \tag{3.24}$$

which may be approximately written

$$\delta x_1^{(i+1)} = \frac{\partial f_1}{\partial x_1} \delta x_1^{(i)} + \frac{\partial f_1}{\partial x_2} \delta x_2^{(i)}$$
$$\delta x_2^{(i+1)} = \frac{\partial f_2}{\partial x_1} \delta x_1^{(i)} + \frac{\partial f_2}{\partial x_2} \delta x_2^{(i)} \tag{3.25}$$

or, more briefly,

$$\delta x^{(i+1)} = J \, \delta x^{(i)} \tag{3.26}$$

where J is the Jacobian matrix of the linearised transformation.

As is known from the theory of linear transformations on the plane, the condition required for contraction (dilation) of the areas is that the modulus of the Jacobian determinant (i.e. corresponding to the matrix J) be < 1 (> 1).

Take, for instance, the linear transformation

$$z = \alpha x + \beta y$$
$$t = \gamma x + \delta y \ .$$

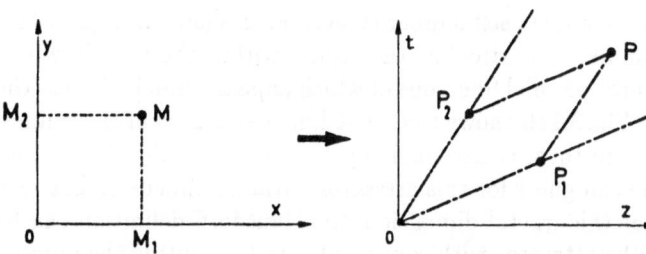

Fig. 3.10. Graphical representation of a linear transformation

From this transformation the unit square OM_1MM_2 (see Fig. 3.10), with $M_1 = (1,0)$, $M_2 = (0,1)$, is transformed into the parallelogram OP_1PP_2, with

$$P_1 = (\alpha, \gamma)$$
$$P_2 = (\beta, \delta) \quad .$$

The area of the latter is given by the modulus of the vector product of OP_1 and OP_2, i.e. by

$$|\alpha\delta - \beta\gamma| \quad .$$

Thus, if

$$|\alpha\delta - \beta\gamma| < 1$$

the areas contract; they dilatate in the opposite case.

Let us now consider a specific example of a two-dimensional map: the so-called "Hénon map" (Schuster, 1984). This map may be considered as the result of a refinement of the logistic map, when the model takes into account the fact that the linear term (the generation term) must refer to the preceding time-lapse.

The Hénon map has the following expression:

$$x_1^{(i+1)} = 1 - a \left(x_1^{(i)} \right)^2 + x_2^{(i)}$$
$$x_2^{(i+1)} = bx_1^{(i)} \quad .$$

(3.27)

The modulus of the Jacobian determinant, in this case, takes the value $|b|$.

Thus, if $|b| < 1$, the map is dissipative. We shall now consider successive iterations of the Hénon map for particular values of the parameters: let us choose, for example, $a = 1.4$ and $b = 0.3$. The result of a large number of iterations is given in Fig. 3.11 (Farmer, 1982). In Fig. 3.11 it can be seen that there is an attractor which is clearly delimited within the plane, as it must necessarily be since $|b| < 1$. Whatever the initial condition within the basin of attraction (whose structure will not be examined here), successive iterations rapidly converge towards the attractor. The sequence of iterations, however, leads to irregular and unforeseeable shifts on the attractor; a chaotic dynamics is then established on the attractor itself. Such an attractor is called a "strange attractor".

The attractor has a complex self-similar structure, as shown in Figs. 3.11a, b and c. Let us examine, in particular, the region within the box drawn in Fig. 3.11a. It shows three parallel lines, one of which appears "thicker" than the others. A close-up in Fig. 3.11b shows that that line is made up of three lines, the thickest of which , in turn, is also made up of three lines (Fig. 3.11c), and so on. Intuitively, one can guess for this attractor a fractal dimension between 1 and 2. To determine this fractal dimension, the Hausdorff definition can be applied, by covering the attractor with squares of side L, counting the number of squares, N, required to completely cover the attractor itself, and going to the limit:

$$D_0 = \lim_{L \to 0} \frac{\ln(N)}{\ln(L)} \quad .$$

A numerical evaluation (with reference to the previously considered values $a = 1.4$, $b = 0.3$) gives $D_0 = 1.26$ (Schuster, 1984).

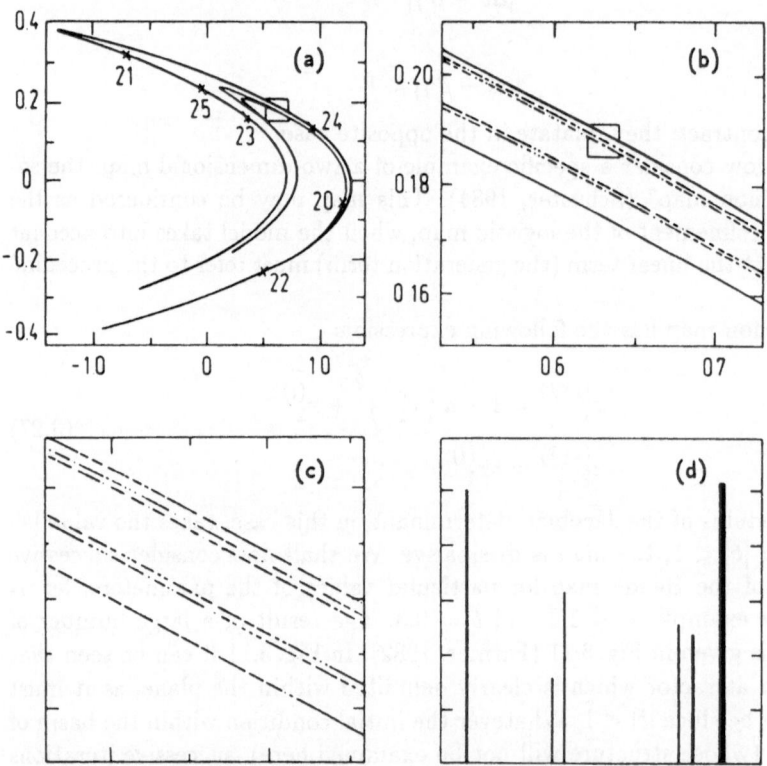

Fig. 3.11. a) The Hénon attractor for 10^4 iterations. Some successive iterates have been numbered to illustrate the erratic movement on the attractor; **b,c)** enlargements of the squares in the preceding figure; **d)** the height of each bar is the relative probability to find a point in one of the six leaves in c) (after Farmer, 1982)

Like in the case of the logistic map, we meet here a chaotic attractor. In this two-dimensional case, hewever, there is both contraction of the areas and sensitive dependence upon the initial conditions (the latter being witnessed by the succession of iterates on the attractor). The explanation of this behaviour lies in the fact that the contraction of the areas is not isotropic. As Fig. 3.12 shows, a set of points forming a circle of a diameter which is small with respect to the dimensions of the attractor is transformed, by the first iteration, into a rather elongated ellipse, and by the second into an almost degenerate ellipse with its main axis parallel to the attractor (Bergé et al., 1984).

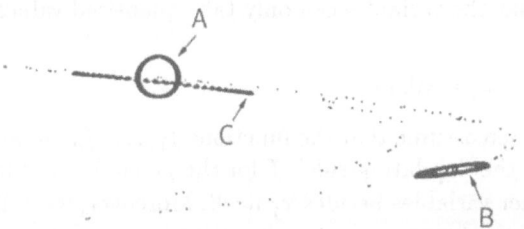

Fig. 3.12. The non-isotropic area deformation by the Hénon attractor (after Bergé et al., 1984)

Generalizing a concept introduced for one-dimensional systems, it can be shown that in this case there are two Lyapunov exponents of opposite sign: the negative one governs the contraction of the minor axis of the ellipse, while the other governs the "stretching" of its major axis.

The definition of Lyapunov exponents for two-dimensional systems is a generalization of that for one dimension. It takes into account the fact that the Lyapunov exponents govern the contraction or the stretching of the distances along two axes which do not necessarily coincide with the coordinate axes.

In the case of the previously considered Hénon system, a numerical evaluation leads to the identification of two Lyapunov exponents which are independent of the initial conditions, of opposite sign and equal to (Vastano and Kostelich, 1986)

$$\lambda_1 = 0.603, \quad \lambda_2 = -2.34 \ .$$

The fundamental concepts for the treatment of systems with many degrees of freedom are the same introduced for two-dimensional systems. Possible attractors are, once again, fixed points, limit cycles or strange attractors. Obviously, the fractal dimension of the latter may be more than 2.

The condition of dissipation in phase space is still given by the modulus of the Jacobian determinant: dissipation occurs if $|J| < 1$. The number of Lyapunov exponents is equal to the dimensionality of the system (Bergé et al., 1986).

3.4 Cellular Automata

Let us consider an N-dimensional autonomous dynamical system:

$$x^{(i+1)} = f(x^{(i)}, P)$$
$$x = \{x_1, x_2, \ldots, x_n\}$$
$$f = \{f_1, f_2, \ldots, f_n\} \tag{3.28}$$
$$P = \{P_1, P_2, \ldots, P_M\}$$

and let us introduce several simplifying hypotheses.

Let us assume, first of all, that the variables can only take quantised values or, more precisely, let

$$x_j \in \{0, 1\}$$

for every j and every i. Let us then assume that the functions $f_1, \ldots, f_n$ are all equal. Let us also assume that the "updating rule" f for the general variable x_j is a function of only two other variables besides x_j itself. Moreover, we will restrict ourselves to systems of the form:

$$x_j^{(i+1)} = f\left(x_{j-1}^{(i)}, x_j^{(i)}, x_{j+1}^{(i)}, P\right), \quad j = 1, 2, \ldots, n \tag{3.29}$$

This kind of systems admits a physical interpretation which is worth describing in some detail, because it is due to this that a system of the type (3.29) can be regarded as a one-dimensional cellular automaton, rather than an n-dimensional one, as the dimension of the state space would suggest.

Let us therefore consider a string of elements ("cells"), each of which can take one of two states (0 or 1) and is connected only to its two nearest neighbours with which it exchanges information about its state. Within this topology, the system (3.29) may be considered as the discrete and quantised analogue of a partial differential equation in a single variable.

In order to completely define the system (3.29), it is necessary to assign a meaning to the variables x_0 and x_{n+1}: the assumptions $x_0 = x_n$ and $x_{n+1} = x_1$ then correspond to the closure into a circle of the one-dimensional string of elements.

It is worth making a consideration of a general nature for these cellular automata (Wolfram, 1984, 1986). Since the single elements (or cells) can only take a limited number of values, automata with a finite number of elements can only assume a finite number of configurations. In particular, a system with n elements has 2^n different possible configurations. Since we are, in fact, dealing with deterministic systems, it can be stated that after a number of time steps equal at most to 2^n the system shows a cyclic behaviour.

When, however, n becomes very large (in the limit $n \to \infty$) the above consideration, while remaining a valid one in principle, becomes of limited interest: more meaningful in this case is the study of the possible periodic behaviours with a period much less than 2^n or of "chaotic" behaviours (i.e., of behaviours without any periodicity with a period less than 2^n), albeit within an inevitable repetition over long periods of time.

In contrast to the systems studied in the previous sections, a general analytical theory of cellular automata is not available: an analysis of their behaviour must therefore be based upon numerical simulation.

Before examining the results about the dynamical behaviour of one-dimensional cellular automata, some considerations should be made about the function f in Eq. (3.29). As mentioned above, this is a boolean function of the states of each cell's nearest neighbours. It is generally a boolean function of three variables, each of which can take either the value 0 or 1; it assigns the value 0 or 1 to each of the combinations of variables (which are $2^3 = 8$, thus giving $2^8 = 256$ possible combinations). If these combinations are arranged as shown in Fig. 3.13, it can be seen that each boolean function can be represented by an eight-figure binary number, or by the corresponding decimal number. This numeration will be used in the figures presented below.

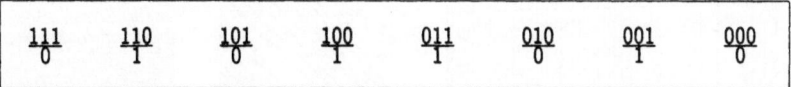

Fig. 3.13. An example of rule for the time evolution of a one-dimensional automaton of the type (3.29). The eight possible states of the three nearest neighbour sites are shown in the upper part of the figure. They are ordered according to the corresponding three-digit binary number. The lower part of the figure shows the "response" to the specific rule, which can be characterized in this case by an eight-digit binary number or by the corresponding decimal number (after Wolfram, 1986)

The dynamical behaviours of cellular automata which correspond to the various possible boolean functions ("rules") differ greatly from one another, as shown in Fig. 3.14, which shows the time evolution of the various possible cellular automata starting from random initial conditions. The different behaviours can be grouped into four classes, according to the type of dependence upon the initial conditions.

Class 1: This class includes those systems which, after some time steps, reach a homogeneous situation which is independent of the initial conditions. The first system of this type in Fig. 3.14 is number 0.

Class 2: This class includes those systems which, after a relatively short initial transient (a few time steps), show, in the space-time plane, simple configurations made up of separate regions which are constant or periodic with period $\ll 2^n$. The first systems of this type in Fig. 3.14 are numbers 1 and 2. These configurations are not independent of the initial conditions. What is relatively independent of the initial conditions, however, is the structure of the configurations. Take, for example, system number 2. The specific position

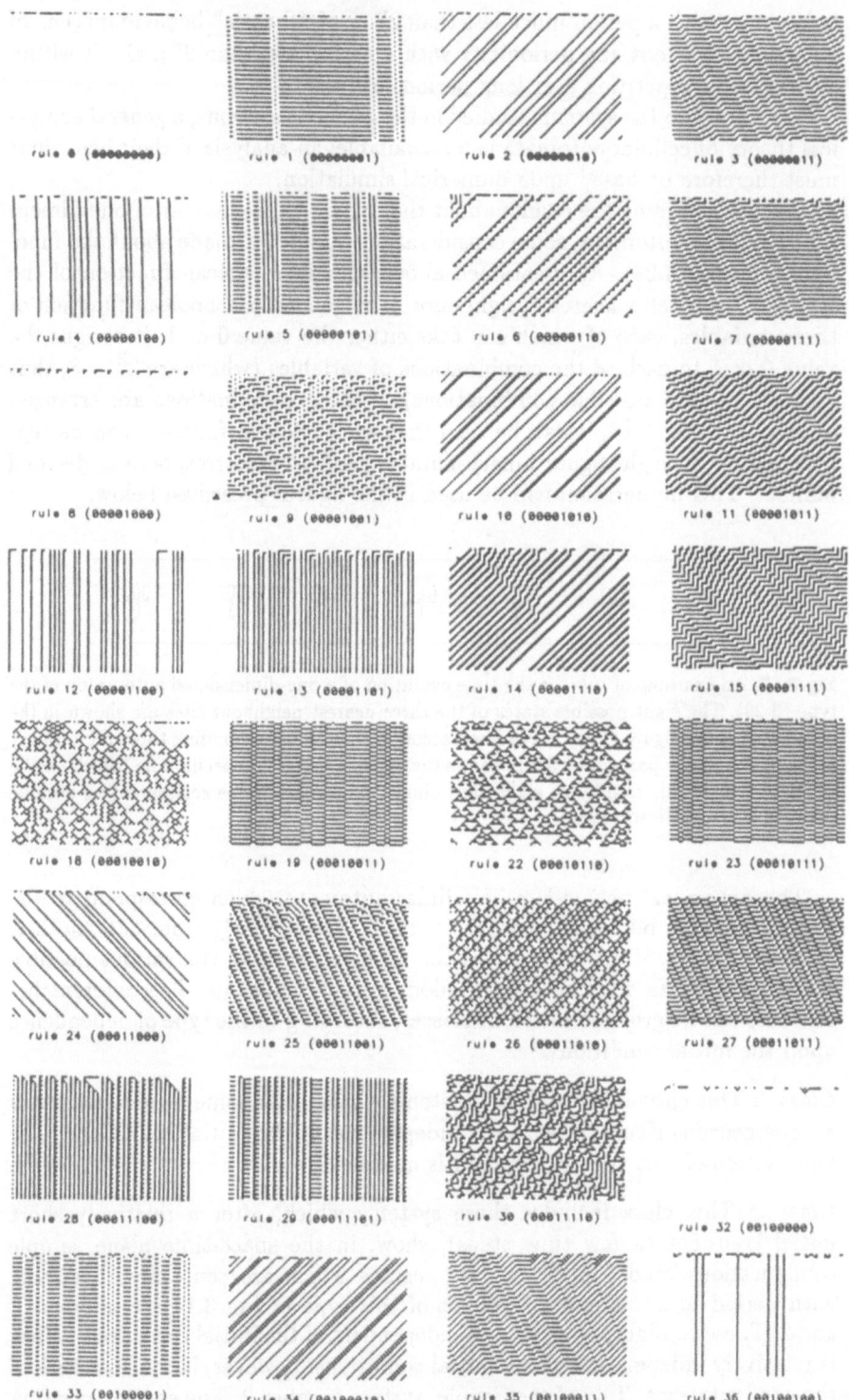

Fig. 3.14. Behaviour of cellular automata of the type (3.29) starting from random initial conditions (80-element automata followed through 60 time steps) (after Wolfram, 1986)

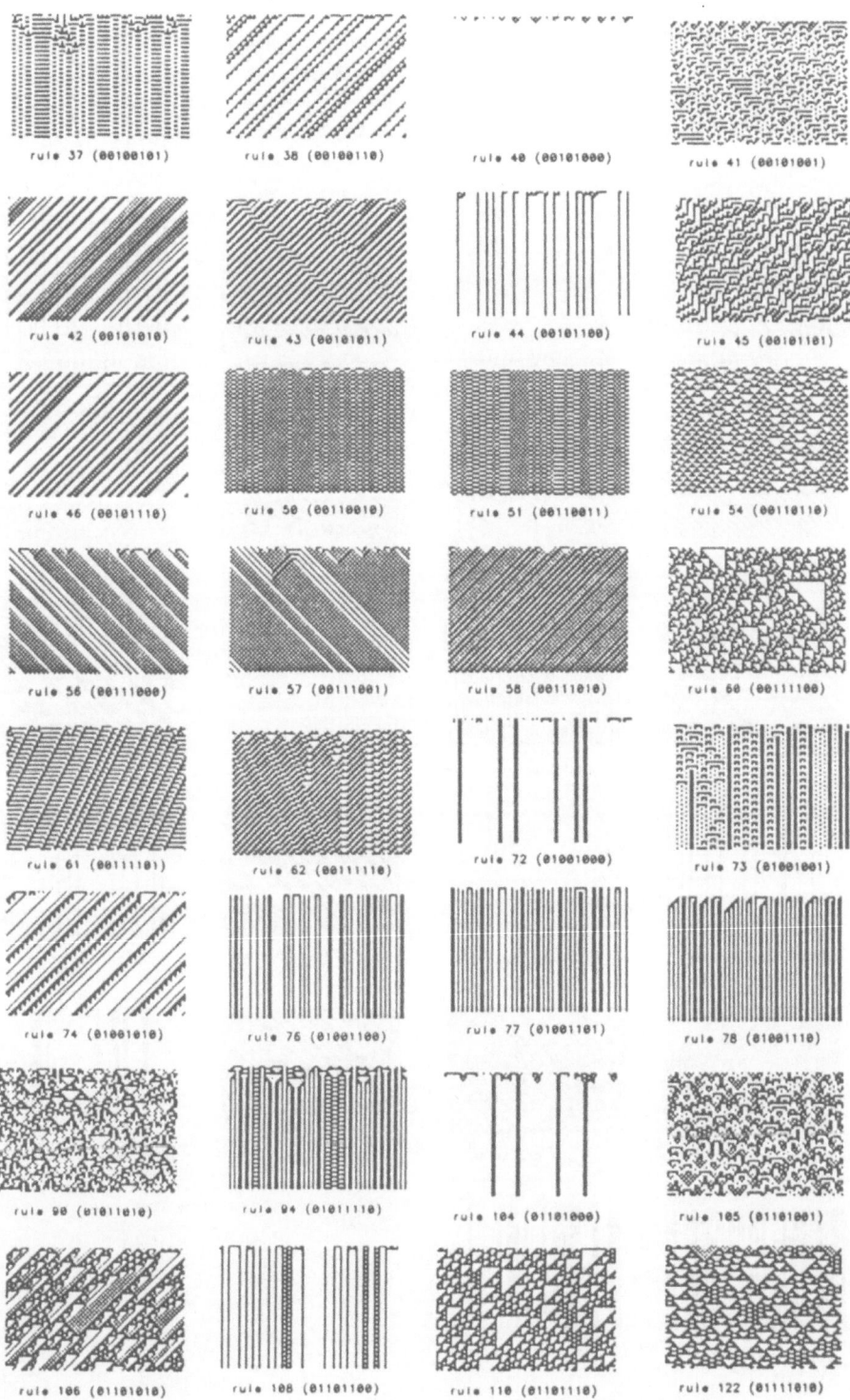

Fig. 3.14 (continued)

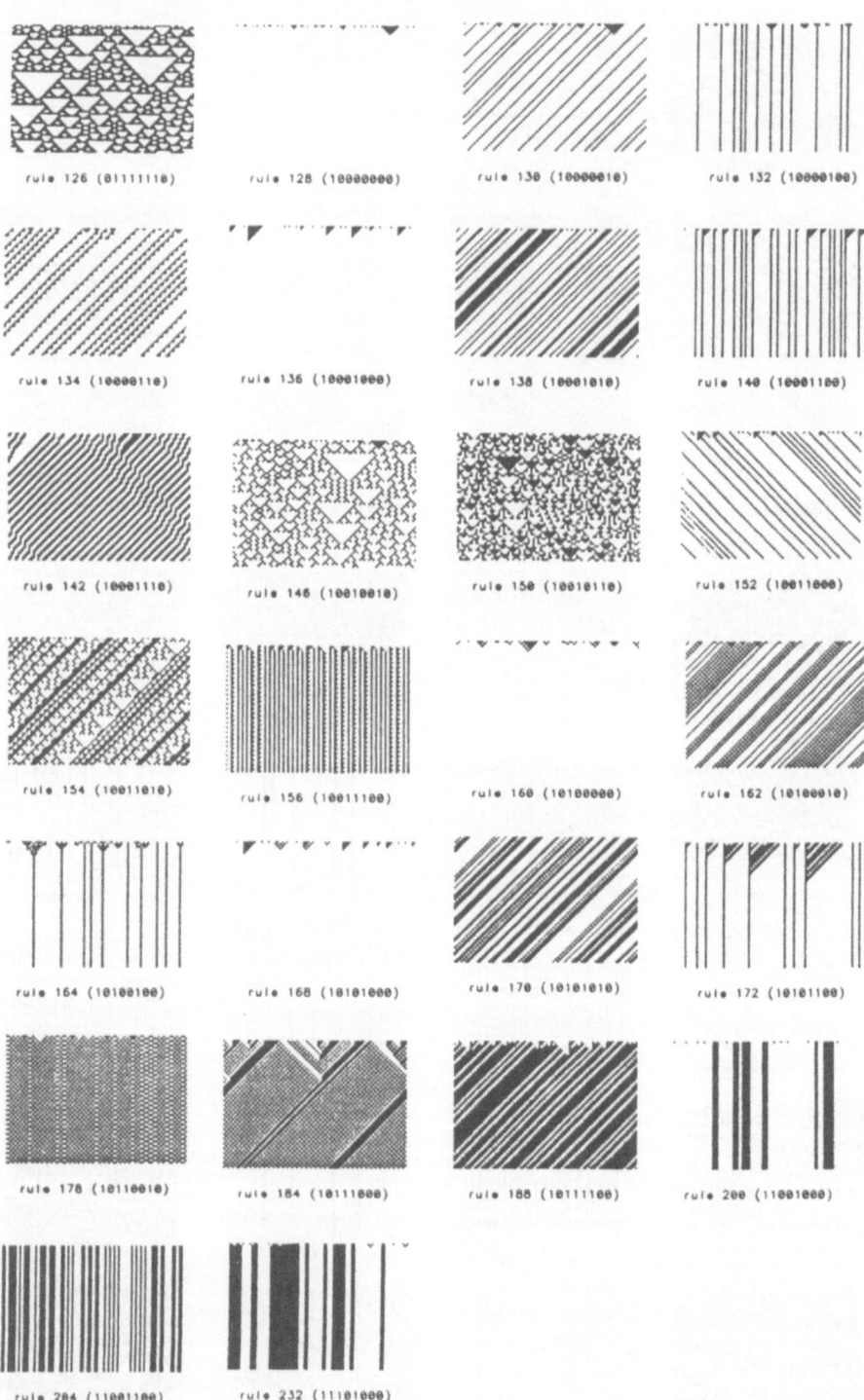

Fig. 3.14 (continued)

of the lines in the space-time plane depends upon particular clusters of values in the initial conditions, whereas the overall behaviour, i.e. the existence of lines in a certain direction in the space-time plane, does not depend upon the initial conditions.

Class 3: This class comprises systems which, starting from random initial conditions, give rise to "chaotic" structures, i.e. to structures which lack any periodicity with a period $\ll 2^n$. The first system of this type in Fig. 3.14 is number 18. It is interesting to examine the behaviour of these systems for initial conditions which are all zero except for a single site (a "seed"). It can be seen that a part of such systems, in this case, behaves like those belonging to the previous classes, while a significant number of these Class 3 systems exhibit a self-similar behaviour. This is illustrated in Fig. 3.15 by the system corresponding to rule 18. When $n \to \infty$, the corresponding space-time pattern has a fractal dimension equal to

$$D_0 = \frac{\ln 3}{\ln 2} = 1.59$$

as can be easily shown by means of the Hausdorff definition.

Class 4: Systems in this class produce highly complex, irregular, localised and propagating structures whose characteristics vary significantly with the initial conditions. In many cases these systems show this kind of behaviour even starting from an initial condition with only one "seed": see, for instance, rule 110 in Fig. 3.15.

These four classes may also be examined from the point of view of their "processing" of the initial conditions. In this light, one may say that the rules corresponding to the first class carry out a trivial processing of the initial conditions. In the second class, the "lines" depend upon small clusters of values in the initial condition: thus, in this case, one has a kind of "recognition" of the structure of the initial conditions. Passing on to the third class, one has an increasing dependence on the initial conditions: for this reason, it is possible to observe chaotic behaviours. In the fourth class, the cellular automata carry out a very complex processing of the initial conditions, such that predictions of a general character are no longer possible.

It can be shown, however, that in these "one-dimensional" systems (which, as mentioned above, are the discrete analogue of partial differential equations with a single quantised space variable) there are only two independent Lyapunov exponents; they measure the velocity of propagation of the information about the initial conditions respectively in one direction and in the opposite along the lattice (Packard, 1985). More precisely, classes 1 and 2 have zero Lyapunov exponents, so that the information about the initial conditions remains localised, and the value of a particular site at any instant in time depends upon the initial values within a fixed interval. Class 3 automata have

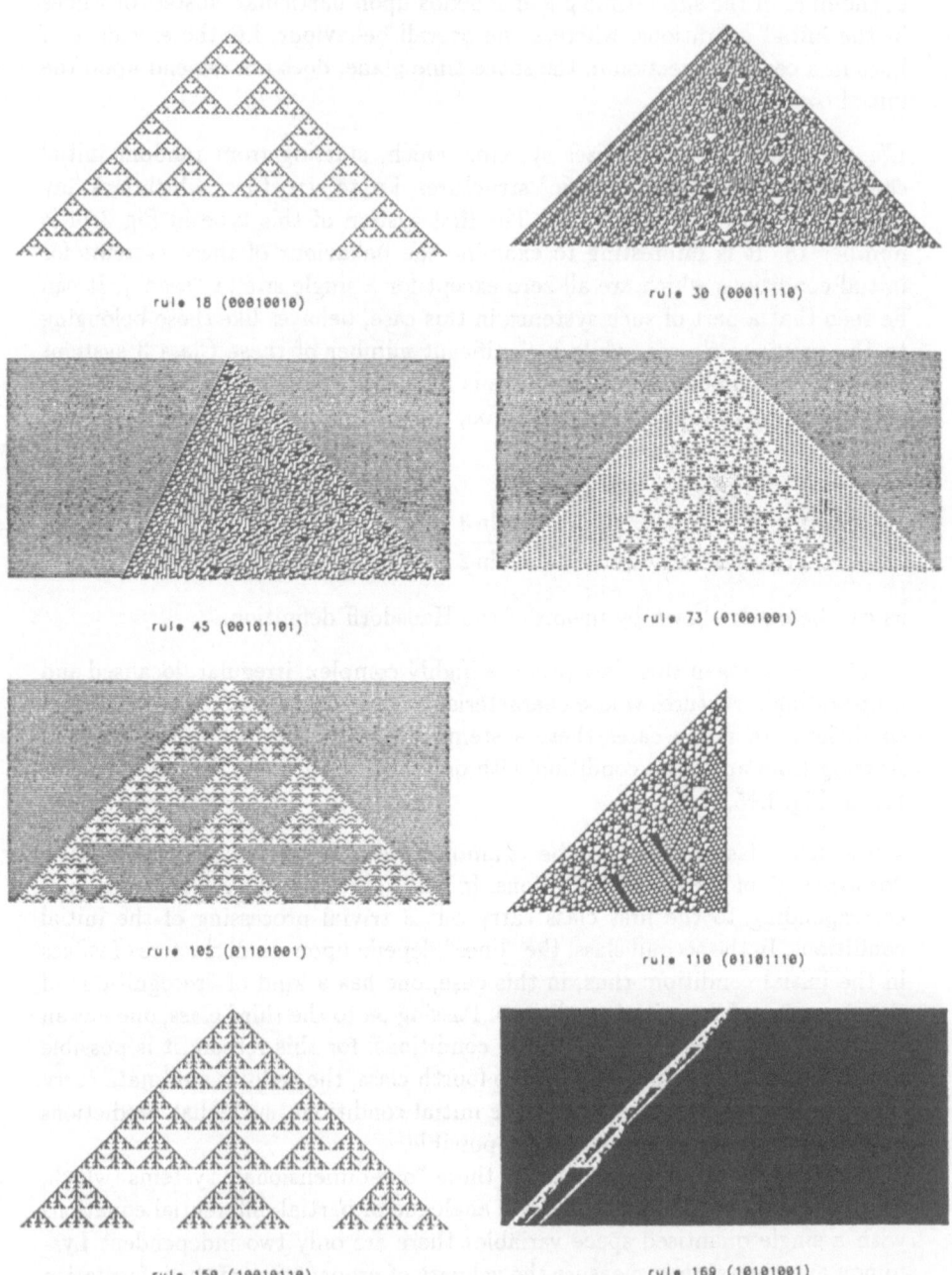

Fig. 3.15. Behaviour of cellular automata of the type (3.29) starting from a single "seed" (100-element automata followed through 100 time steps) (after Wolfram, 1986)

positive Lyapunov exponents, so that a small difference in the initial conditions diffuses over the whole lattice, and the value of a particular site after a sufficiently large number of time steps depends critically upon the initial values. Finally, class 4 automata have positive Lyapunov exponents which, however, tend asymptotically towards zero (Packard, 1985).

If we consider, in particular, class 2 systems, we can say that they "recognise" specific spatial patterns within the initial conditions. The problem of pattern recognition by one-dimensional cellular automata has been studied in detail by Jen (Jen, 1986), who presented theoretical results on "filtering rules" and on the minimum neighbourhood size of the rules. In particular, it has been shown that there is a minimum neighbourhood radius for every cell (i.e., a minimum number of neighbours on which values any cell depends) which can allow the recognition of a given class of patterns.

These theoretical results will not be discussed in detail here: it is worth recalling them, however, because they enlarge the range of cognitive tasks which simple one-dimensional cellular automata are capable of carrying out.

Figure 3.16 shows the result of applying filtering rules with different neighbourhood sizes to the recognition of the pattern 011111 in the initial conditions. It can be seen that the execution of this cognitive task requires a neighbourhood radius not less than 5 (which implies 10 neighbours for each cell).

Two-dimensional cellular automata are particular examples of Eq. (3.28) where the rhs does not depend upon all the state variables but, for each element x_j of the state vector, only upon those elements which are within a suitable neighbourhood. Such a neighbourhood, however, can be no longer represented on a line, as in the case of Eq. (3.29), but rather on a plane. In other words, in two-dimensional cellular automata, the topology of the connections between elements allows the system to be interpreted as a discrete analogue of a partial differential equation with two space variables.

Two-dimensional cellular automata may be based upon various lattice structures and neighbourhood dimensions. "Moore" neighbourhoods, which consist of 9 sites (the site itself and its eight nearest neighbours) are very often adopted.

In this case there are 2^{512} possible rules, and an exhaustive treatment is then impossibile. Fig. 3.17 provides some examples of structures generated after 24 time-lapses starting from a single "seed" with rules which have different symmetry properties, although they all belong to the "nine neighbourhood" class (Wolfram, 1986). In the absence of symmetry, also "dendritic" growth is possible: an example of this behaviour, starting from a small disordered region, is shown in Fig. 3.18 for different time values.

In most two-dimensional cellular automata, there is a strong dependence upon the initial conditions. According to the initial conditions, extremely different behaviours (time-independent, periodic, chaotic, etc.) may occur for

```
pattern to be recognized:    011111

input sequence:
0111010110111111010111001110001110111110101101111110000010011010010110000

radius of rule used:    2

evolution of automaton:
0111010110111111010111001110001110111110101101111110000010011010010110000
01 1 0  0111111  01 1 01 1  01 1011111  0  0111111
        0111111            011111       0111111
        0111111            011111       0111111
        0111111            011111       0111111
        0111111            011111       0111111
        0111111            011111       0111111

radius of rule used:    3

evolution of automaton:
0111010110111111010111001110001110111110101101111110000010011010010110000
0        0111111  0   0   0   011111       0111111
         0111111              011111       0111111
         0111111              011111       0111111
         0111111              011111       0111111
         0111111              011111       0111111

radius of rule used:    4

evolution of automaton:
0111010110111111010111001110001110111110101101111110000010011010010110000
         0111111              011111       0111111
         0111111              011111       0111111
         0111111              011111       0111111
         0111111              011111       0111111
         0111111              011111       0111111
         0111111              011111       0111111

radius of rule used:    5

evolution of automaton:
0111010110111111010111001110001110111110101101111110000010011010010110000
         011111               011111       011111
         011111               011111       011111
         011111               011111       011111
         011111               011111       011111
         011111               011111       011111
         011111               011111       011111
```

Fig. 3.16. One-dimensional cellular automata confronted with the cognitive problem of recognizing the sequence 011111 (after Jen, 1986)

the same rule. In order to illustrate this behavioural variety with a particular two dimensional cellular automaton, the next section will briefly examine what is probably the most famous example of such systems: Conway's "life game".

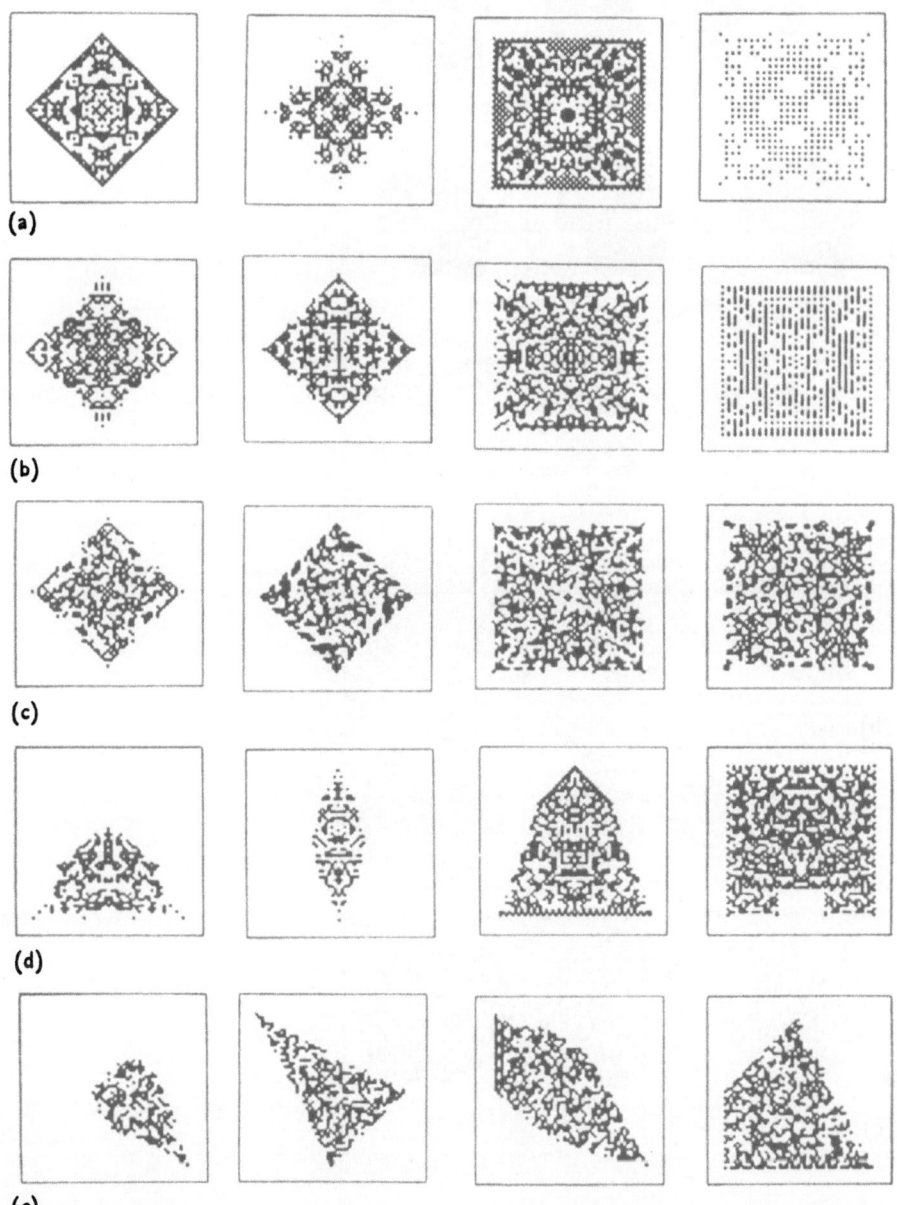

Fig. 3.17a-e. Examples of patterns generated by the evolution of two-dimensional cellular automata starting from a single "seed". Row a) rules with a complete symmetry; row b) rules with horizontal and vertical reflection symmetry; row c) rules with rotation symmetry; row d) rules with vertical reflection symmetry; row e) rules without symmetry (after Wolfram, 1986)

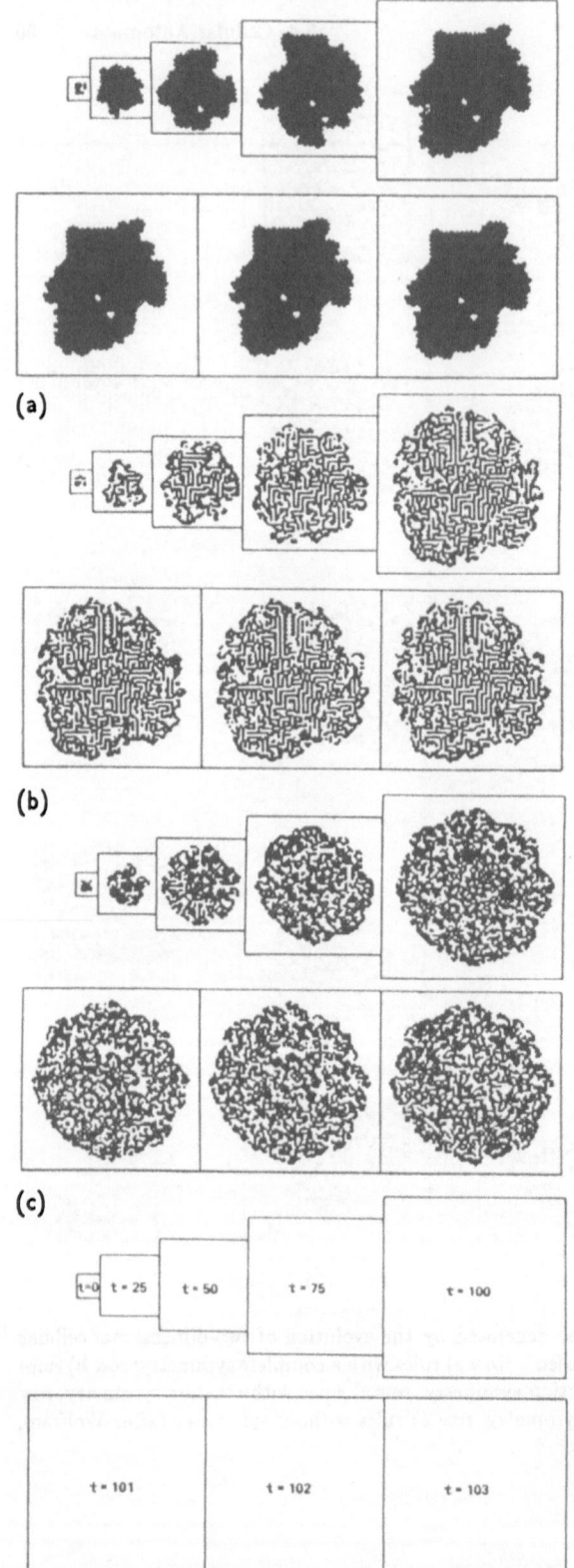

Fig. 3.18. Examples of two-dimensional cellular automata showing a "dendritic" growth starting from small disordered regions (after Wolfram, 1986)

3.5 The Life Game

The life game, proposed by John Conway (Gardner, 1983; Poundstone, 1985), is a particular case of a two-dimensional cellular automaton. In it, the variables are boolean and the functions f_i defined by Eq. (3.28) are, as above, all equal: they are boolean functions of the values of the states of the variables contained within a Moore neighbourhood.

The definition of the boolean function is inspired by a socio-biological metaphor (from which the name of the game derives): the hypothesis is that there exists an optimal population density for survival. At each time, for every variable (or "cell"), the number of nearest neighbours (the cell itself being excluded) in the "on" condition is computed. After one time-lapse (in the "next generation"), the cell being considered will then:

- remain in the same state if there are two nearest neighbours in the "on" condition;
- be in the "on" condition, whatever its previous condition, if it has three nearest neighbours in the "on" condition;
- be in the "off" condition in any other case (i.e., if its nearest neighbours in the "on" condition are zero, one, four, five, six, seven or eight).

In order to understand how such a simple rule can give rise to extremely complex behaviours, a brief analysis will be made, starting from different initial conditions.

First of all, let us consider the case where all the cells are "off". The rule states that, in this case, the cells will remain "off" for all successive time steps. If, as the initial condition, we take the opposite situation (that is, we suppose all the cells "on"), the rule states that, also in this case, the next step will "switch off" all the cells. The same transition from "on" to "off" occurs both for isolated cells and for adjacent cell pairs which are "on".

Therefore, the first non-trivial behaviour occurs when there are three aligned "on" cells. The resulting behaviour, as shown in Fig. 3.19a, is an oscillation with period 2: this object is often called a "blinker". The other possible triplet of "on" cells, the "L triplet", leads, after only one time lapse, to a 2×2 square (a "block") which is time independent (Fig. 3.19b): this is the first example of a stationary configuration which is referred to as "still life" (Poundstone, 1985).

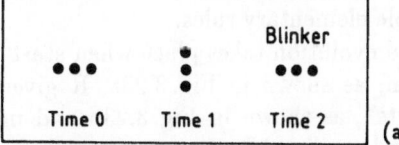
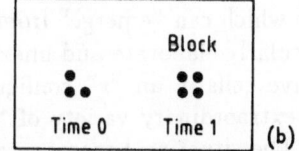

Fig. 3.19a,b. Time evolution of simple initial configurations of "life game". a) Three aligned "on" cells; b) three "L" shaped "on" cells

Let us go on to consider initial configurations with more than three "on" cells. The initial condition with four aligned "on" cells leads, after only two time lapses, to a still life structure: a hexagon of six cells (or "beehive", see Fig. 3.20). The "T" structure with four cells "on", after 10 time lapses, leads to a "traffic-light" structure, oscillating between the configurations of step 9 and 10 with period 2 (Fig. 3.21).

Time 0   Time 1   Time 2

Fig. 3.20. Time evolution of an initial configuration with 4 aligned "on" cells

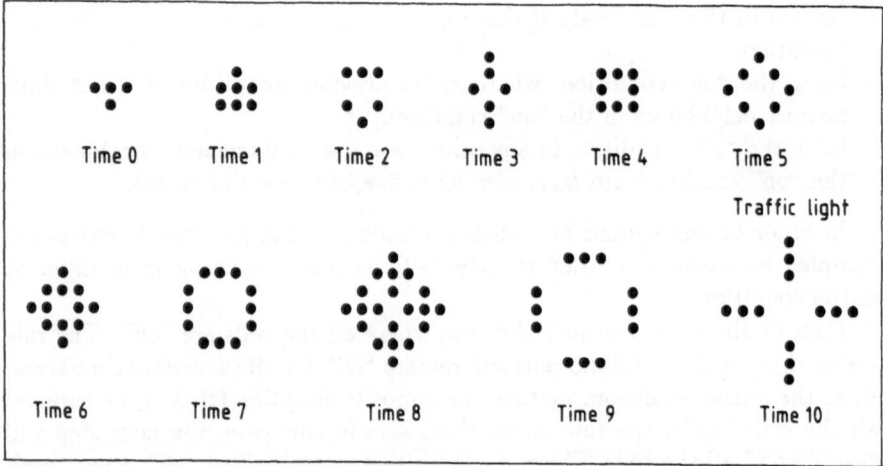

Fig. 3.21. Time evolution of an initial configuration with 4 "T" shaped "on" cells

The configurations with five cells "on" give rise to quite different behaviours. Whereas a string of five aligned cells evolves in eight steps to a "traffic-light", the "glider" in Fig. 3.22 shows an unexpected behaviour. It moves diagonally across the plane at a constant speed (a diagonal square every fourth time-lapse), changing its shape with period 4, unless it meets other "objects". The "glider", which is frequently formed when starting from random initial conditions, is a typical example of the complex and unexpected behaviours which can "emerge" from simple elementary rules.

A particularly elaborate and unexpected evolution takes place when starting from five cells in an "r" configuration, as shown in Fig. 3.23a. It gives rise to an extraordinary variety of "objects", as shown in Fig. 3.23b and in Fig. 3.23c. The situation becomes stable and predictable after over one thousand time steps: the configuration then consists of 6 "gliders", 15 still life objects of various dimensions and 4 "blinkers" (Poundstone, 1985).

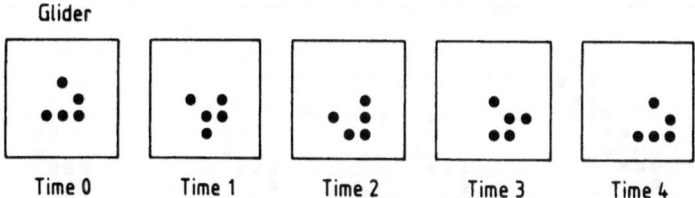

Fig. 3.22. Time evolution of an initial configuration with 5 "glider" shaped "on" cells

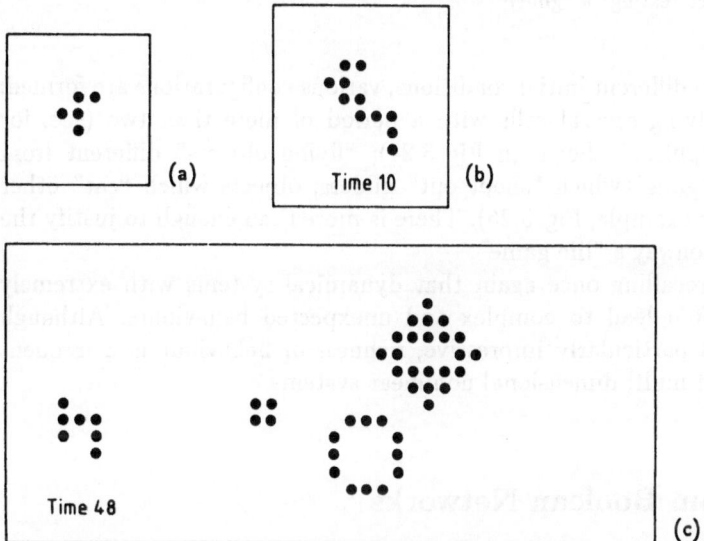

Fig. 3.23a–c. Time evolution of an initial configuration with 5 "r" shaped "on" cells. a) initial condition; b) configuration at $t = 10$; c) configuration at $t = 48$ (after Poundstone, 1985)

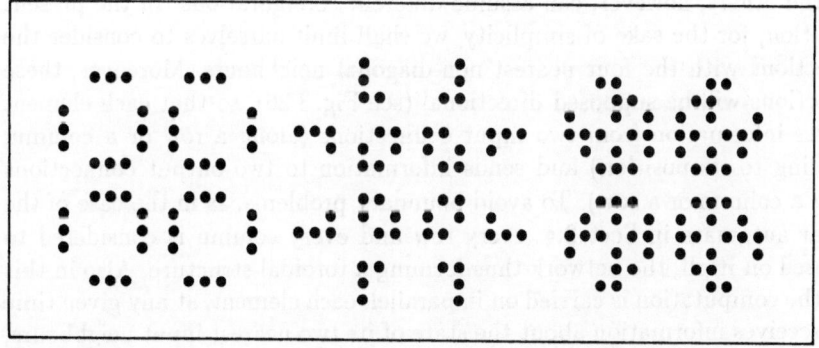

Fig. 3.24. An example of oscillator with period 3 composed by many elements

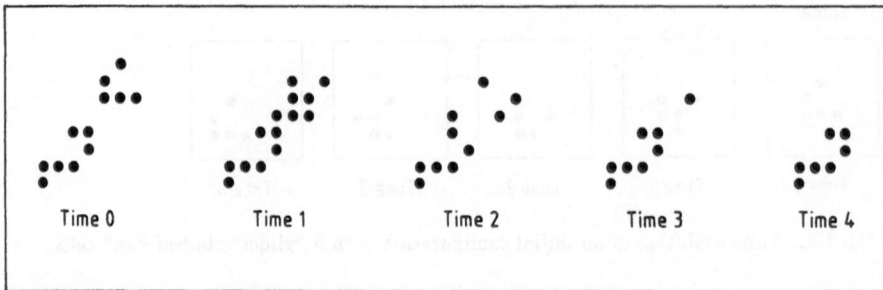

Fig. 3.25. An object "eating" a "glider"

Starting from different initial conditions, various configurations are formed: oscillators involving several cells with a period of more than two (see, for example, the "pulsar" shown in Fig. 3.24); "flying objects" different from the "gliders"; "guns" which "shoot out" gliders; objects which "eat" other objects (see, for example, Fig. 3.25). There is more than enough to justify the popularity of Conway's "life game".

It is worth recalling once again that dynamical systems with extremely simple rules often lead to complex and unexpected behaviours. Although this example is particularly impressive, richness of behaviour is a frequent characteristic of multi-dimensional nonlinear systems.

3.6 Random Boolean Networks

Another class of two-dimensional cellular automata which has been studied in detail by many authors is provided by random boolean networks (Kauffman, 1984; Atlan et al., 1981).

A network of this type, like the life game, consists of elements which can take on one of two possible states, 0 or 1. The connections of each element with the others, however, can assume different configurations. In the present exposition, for the sake of simplicity, we shall limit ourselves to consider the connections with the four nearest non-diagonal neighbours. Moreover, these connections will be supposed directional (see Fig. 3.26), so that each element receives information from two input connections (along a row or a column, according to its position) and sends information to two output connections (along a column or a row). To avoid boundary problems, as in the case of the cellular automata in Sect. 3.4, every row and every column is considered to be closed on itself, the network thus forming a toroidal structure. Also in this case, the computation is carried on in parallel: each element, at any given time step, receives information about the state of its two nearest input neighbours, redefines its own state and, at the next time step, sends this information to its two nearest output neighbours.

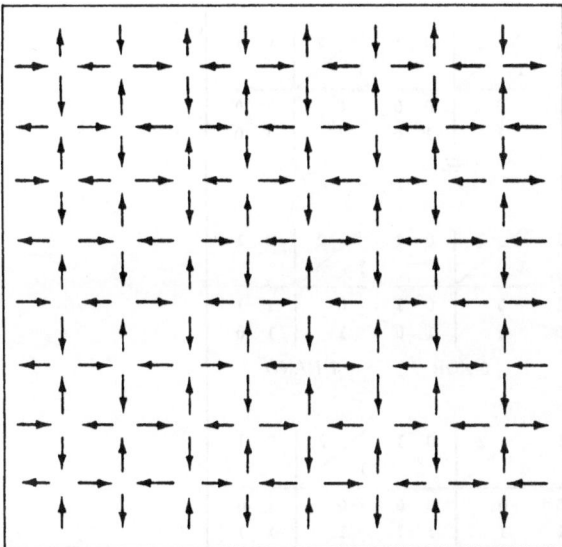

Fig. 3.26. The connections in a standard random boolean network

The fundamental difference between random boolean networks and cellular automata lies in the fact that the rule according to which the elements redefine their own state is, in random boolean networks, not homogeneous, but can vary from element to element. More precisely, each element carries out its computation according to one of the sixteen possible boolean functions of its two inputs (see Fig. 3.27). Let us suppose, for example, that the boolean functions are defined at the beginning of each run through a random assignment procedure and are then left unchanged. Fig. 3.28 shows an example of a 16×16 network with a random distribution of the boolean functions of Fig. 3.27.

The interest in the study of this type of network was born in a biological environment. In fact, random boolean networks were first introduced by Kauffman (Kauffman, 1970) as abstract models for gene regulation mechanisms. This reference is particularly interesting, because it exemplifies quite an important line of approach towards the dynamics of complex systems.

Kauffman studied the problem of finding the collective properties to be expected from a system of genes interacting through repression-excitation mechanisms. His aim was not to describe a particular organism, but to study the "generic" properties of an entire class of systems (such as, for example, their stability), in order to understand the mechanisms of biological evolution.

The representation chosen was therefore an abstract and greatly simplified description of the interaction mechanisms, because the emphasis was laid upon collective properties.

One might question the correctness of this approach, which is based upon the hypothesis of the existence of overall properties which are largely inde-

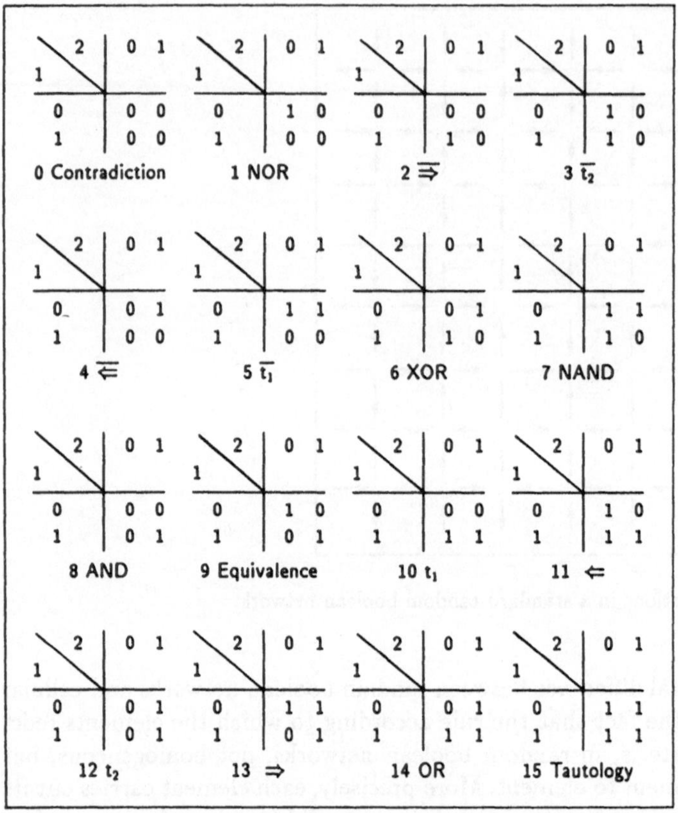

	0	1
0	0	0
1	0	0

0 Contradiction

	0	1
0	1	0
1	0	0

1 NOR

	0	1
0	0	0
1	1	0

2 $\overline{\Rightarrow}$

	0	1
0	1	0
1	1	0

3 $\overline{t_2}$

	0	1
0	0	1
1	0	0

4 $\overline{\Leftarrow}$

	0	1
0	1	1
1	0	0

5 $\overline{t_1}$

	0	1
0	0	1
1	1	0

6 XOR

	0	1
0	1	1
1	1	0

7 NAND

	0	1
0	0	0
1	0	1

8 AND

	0	1
0	1	0
1	0	1

9 Equivalence

	0	1
0	0	0
1	1	1

10 t_1

	0	1
0	1	0
1	1	1

11 $\Leftarrow$

	0	1
0	0	1
1	0	1

12 t_2

	0	1
0	1	1
1	0	1

13 $\Rightarrow$

	0	1
0	0	1
1	1	1

14 OR

	0	1
0	1	1
1	1	1

15 Tautology

Fig. 3.27. The sixteen binary logic functions of two inputs

1	3	8	13	3	14	13	12	3	8	10	11	9	14	12	2
11	3	12	13	7	2	5	2	14	6	9	2	7	1	6	7
4	7	8	6	13	4	12	14	9	13	1	3	9	5	14	5
4	13	3	5	2	13	12	8	4	1	9	11	1	14	7	5
12	12	10	8	12	13	12	1	10	8	6	14	3	3	14	4
12	5	6	12	10	12	12	10	3	13	2	9	13	8	6	10
4	10	2	9	2	13	7	3	3	11	9	3	5	13	8	8
10	4	7	13	13	14	1	14	7	5	6	13	5	11	8	9
8	13	8	10	1	3	3	11	8	2	8	14	11	11	9	3
4	8	10	12	8	2	14	4	13	7	9	14	3	3	13	14
6	10	13	5	14	7	4	10	13	7	9	12	13	4	10	10
2	9	13	7	13	6	5	6	14	5	9	11	6	3	9	9
10	14	9	3	3	13	8	2	14	13	1	6	9	10	2	6
5	8	12	7	9	5	9	7	8	3	8	10	14	9	11	12
2	12	13	6	11	7	13	5	11	6	11	12	5	4	1	2
10	6	11	4	8	5	4	5	13	3	3	9	10	3	7	8

Fig. 3.28. An example of 16 × 16 boolean network with a random distribution of logic function, numbered as in Fig. 3.27

pendent of the detailed nature of elementary interactions. There can be no a priori guarantee of this independence; it must be noted, however, that many complex physical systems exhibit this kind of properties with an astounding precision. For example, in the study of phase transitions there are critical exponents which are identical, to many decimal places, for widely different physical systems such as, for instance, liquid crystals, solution polymers, liquefaction processes (Haken, 1978). The existence of generic properties independent of the detailed nature of the interactions suggests that this kind of investigation, oriented towards the search for such properties in simplified models, is both important and promising.

Random boolean networks exhibit marked time self-organization properties (Atlan, 1985). In particular, after a transient of no more than several hundred iterations, they give rise to periodic behaviours with periods which are very short (several tens of time steps) with respect to the maximum allowable period for a network with a finite number of states, and are independent of the initial conditions. On increasing the number of elements, N, in the network, the cycle period increases as $N^{1/2}$ (Atlan et al., 1981).

Besides this temporal self-organization, random boolean networks exhibit spatial self-organization. In particular, there are sub-networks of oscillating elements in a "sea" of stationary ones (i.e., those which have a time-independent state). The period of the whole network is then the lowest common multiple of the cycle periods of the various oscillating sub-networks (Atlan, 1987).

Figure 3.29 shows four different examples of asymptotic behaviour (corresponding to randomly defined different initial conditions) of the network in Fig. 3.28 (Atlan, 1987). It can be seen that some parts of the network are insensitive to changes in the initial conditions: they constitute "cores" which are stationary, or oscillate. The remaining elements modify their behaviour as a function of the initial conditions.

To better understand the space-time self-organization properties of these networks, a slightly more detailed analysis of the role played by the various boolean functions is useful.

Let us, first of all, consider rule 0 ("contradiction") and rule 15 ("tautology"). They generate an output which is independent of the two inputs, and thus play the role of a "filter" with respect to the oscillations of the input elements. One might think that the formation of the stationary sub-networks could be due to the presence of such functions. This hypothesis, however, is contradicted by numerical simulations: in fact, if these two functions are excluded, there is only a slight lengthening of the periods of oscillation.

All of the remaining boolean functions, excluding rule 6 ("exclusive OR") and rule 9 ("equivalence"), are "forcing" or "channelling" structures, in the sense that at least one of their inputs has a value which guarantees the presence of a particular output value. An increase in the relative number of rule 6 and rule 9 units in a network systematically leads to an increase in the length of the cycles, so demonstrating that self-organization properties are essentially connected to the characteristics of the forcing structures. The latter, in fact, channel particular values from one element to another, ignoring the contribution from the second input and thus allowing the "crystallization" of forcing structures in particular sub-networks.

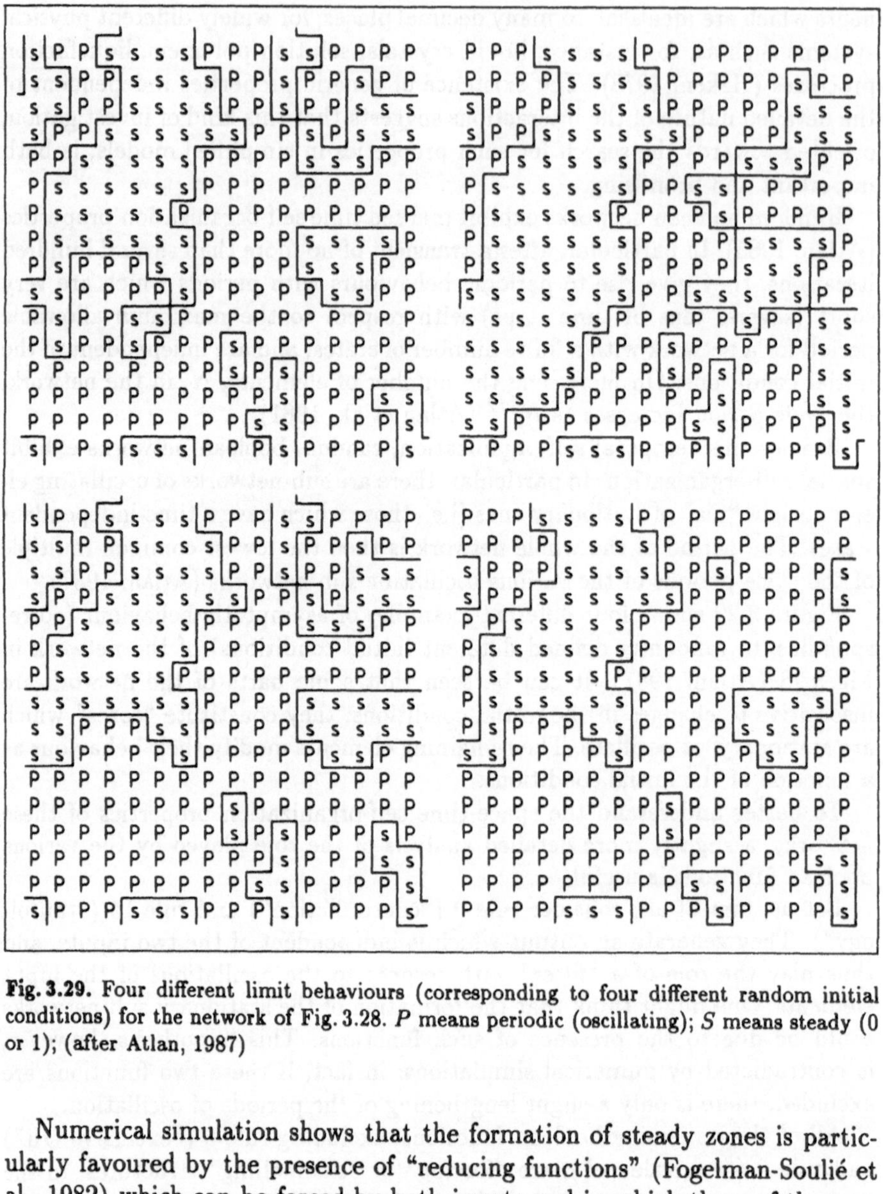

Fig. 3.29. Four different limit behaviours (corresponding to four different random initial conditions) for the network of Fig. 3.28. *P* means periodic (oscillating); *S* means steady (0 or 1); (after Atlan, 1987)

Numerical simulation shows that the formation of steady zones is particularly favoured by the presence of "reducing functions" (Fogelman-Soulié et al., 1982) which can be forced by both inputs and in which three of the possible outputs are identical (functions 1, 2, 4, 7, 8, 11, 13, 14). The remaining rules (3, 5, 10, 12) have the characteristic of only being forceable by one of their inputs, which is the only one effective in determining the output value. In the presence of functions of this type ("transfer functions") the network then decomposes into closed circuits and "branches". In networks composed only of transfer functions, the formation of stable structures is statistically unfavoured and most of the network elements oscillate (Atlan et al., 1981).

The stability of the limit configurations of random boolean networks can be studied by introducing transient perturbations into one or more elements of the system and then allowing it to relax again towards an asymptotic state. It can be seen that, if the perturbation does not extend over too large a part of the network, the network itself usually relaxes to a situation "not too far" from the preceding one. It is thus possible to identify sorts of "basins of attraction" corresponding to various patterns of initial conditions (Atlan et al., 1981). In this case, the network may be considered, in a sense, as an "associative memory": starting from initial conditions belonging to a certain basin of attraction, it tends in fact towards the corresponding stationary configuration.

The overall self-organization of the network into functionally independent zones (sub-networks) is clearly seen by comparing the behaviour of a part of a network when it is included in the overall network and when it is considered as a separate network. The oscillating "patterns" are essentially the same (apart from regions very close to the boundaries).

The structuring of random boolean networks into stable and oscillating regions shows a remarkable resistance to localized transient perturbations. In other words, when a perturbation is applied and then removed, these networks tend to evolve towards an asymptotic state which is very similar to the unperturbed one.

When, on the contrary, a perturbation is permanently applied to a random boolean network, it can induce modifications in the subdivision between stable and oscillating regions.

A first effect is the transformation of stable elements into oscillating ones, through a chain of forcing bounds from the input elements. A second possibile effect is, on the contrary, the transformation of oscillating elements into stable ones (Atlan, 1987).

Moreover, it is interesting to quote the fact that the cyclic introduction of particular sequences of perturbations in a point of a random boolean network can change its asymptotic configuration. This fact can be described as a sort of "recognition" of these sequences by the network, through a self-organizing behaviour. Following Atlan (Atlan, 1987), we can see this dynamical process as an elementary form of spontaneous creation of meaning in an artificial cognitive system.

Recently, a method has been proposed for using these self-organization characteristics of random boolean networks to carry out externally defined cognitive tasks (Paternello and Carnevali, 1987). In particular, it has been shown that a random boolean network can self-organize to reproduce the behaviour of an 8-bit binary adder, when exposed only to a small subset of all possible additions of 8-bit operands.The proposed method of learning by examples refers to an energy function and to a process of "simulated annealing", as in the learning method of the "Boltzmann machine", which will be discussed in Chap. 5.

3.7 Computation in Reaction-Diffusion Systems

The dynamical systems presented so far were intentionally very simple ones. In particular, even when a precise topological structure was referred to, the assumption of cells with a single boolean state variable was maintained. If this assumption is relaxed, and a larger number of variables with varying strength are allowed for each cell, not only can more complex behaviour be observed, but, above all, new perspectives can be opened up for cognitive applications.

The reaction-diffusion models proposed by Steels (Steels, 1987, 1988), which will be briefly outlined below, lie in this class. Their name derives from a mathematical-physical analogy. Every state variable, in fact, can be characterised by its dynamics of propagation from one cell to another, which in turn can be specified using the discrete analogue of a diffusion coefficient. Moreover, the value of each state variable may depend upon the values of the others. For example, the fact that a certain variable exceeds a pre-determined intensity may activate or inhibit another variable, through the analogue of a reaction mechanism.

It must be recalled, at this point, that in chemical systems with reaction and diffusion interesting and even spectacular self-organization behaviours have been observed (Haken, 1980). It can therefore also be expected that discrete reaction-diffusion systems are likely to present a wide range of dynamical behaviours.

Going back for a moment to the life game, introduced in Sect. 3.5, one can immediately realise the greater complexity which could be reached with the changes now proposed. For example, the possibility of having more state variables could allow the representation, on the spatial lattice, of various forms of life instead of only one.

It would become possible, for instance, to model the space-time evolution of the relational dynamics between a foreign substance and an antibody and thus to simulate, in particular, an immunological system-like behaviour (Steels, 1987). The presence of antibodies, in fact, is enforced by the foreign substance; the antibodies move towards the region where the foreign substance is present; the reaction between the two components then leads, locally, to the destruction of the intruder. Clearly, this behaviour can be represented through suitable elementary rules assigned to the cells of a network, if these cells have a sufficiently large number of state variables.

To exemplify the possibilities of reaction-diffusion models, we shall refer to a problem studied by Steels (Steels, 1988), which opens up interesting possibilities in the field of cognitive processes: more precisely, in the domain of problem solving processes.

Let us consider a simple robot, which can move over an orthogonal plane lattice (such as the one in life game) by a unit length in one of the fundamental directions at each time step. The externally assigned task (the goal) is that of reaching a given position. Let us then assume that other objects are placed on the lattice and that they behave as unmoveable obstacles on the way of reaching the goal. The problem is then to identify the state variables and the rules necessary for the robot to "organize itself" to reach the goal, in the absence of a higher level representation of this problem-solving task.

The following state variables can be used:

Robot position (equal to zero everywhere, except at the point where the robot is);

Pull (a quantity diffused from the goal location, which provides the robot with an information upon the direction to be taken).

The rules necessary for the robot to successfully solve the problem are as follows:

Pull diffusion, ruled by a suitable diffusion constant (with a cancellation at the sites of the obstacles);

Definition of the move to be made (towards the cell belonging to the Moore neighbourhood of the robot where there is the highest positive strength of the variable *Pull*).

As it can be seen, a physical analogy has been adopted with the emission of a signal from the goal (a sort of "smoke signal"). Steels has shown that a system of this type effectively works in the case considered (Steels, 1988).

It is also possible to extend the rules and the state variables in order to face up successfully more complex situations (which involve, for instance, the possibility that the robot removes obstacles from his path).

The robot problem is a prototype of a class of cognitive problems which may be tackled by a reaction-diffusion system. More specifically, they are "common-sense problem solving" situations (i.e., without higher level representations of the problem) based upon spatial reasoning.

A particularly impressive example is that of puzzles, classical test benches for artificial intelligence, which were also solved by Steels using dynamical systems (Steels, 1988).

4. Homogeneous Neural Networks

4.1 Introduction

The dynamical system which will be discussed in this chapter was introduced by Hopfield (1982), who relied also upon previous suggestions and results by other researchers (Hebb, 1949; Cooper, 1973; Kohonen, 1972; Little, 1974). As we shall see, the cognitive capabilities of this model are subject to severe limitations. Its clear and elegant mathematical formulation, however, renders it particularly useful for illustrating and examining in detail the main concepts introduced in the preceding chapters.

Moreover, Hopfield's formulation, introducing an energy function, has also shown a relationship between neural networks and the physics of disordered materials, thus opening up the possibility of a dialogue between two scientific communities which previously had no cultural interchange.

The Hopfield model also holds a significant historical importance because its introduction coincided with the renaissance of a widespread interest in neural networks. As we have already mentioned, neural networks had become a quite marginal research topic in the 70's. The recent inversion of this trend can be attributed to the combination of various factors, as discussed in Chap. 2.

One of these factors is precisely the presentation of the Hopfield model, which raised a great interest in the physics community. Several researchers in statistical mechanics then became interested in cognitive systems, and this interest propagated towards researchers in the dynamics of complex systems. Although the dynamics of the original Hopfield model is simple (its attractors being fixed points only), one can develop generalizations of the model which show much more varied behaviours (Sompolinsky and Kanter, 1986; Sompolinsky et al., 1988).

Before examining the behaviour of the Hopfield model in the execution of cognitive tasks, another class of applications of the Hopfield model will be mentioned: optimization problems. Although a detailed analysis of this aspect would lead us away from the main theme of this book, it deserves a brief examination.

As we shall see, the Hopfield model is a dynamical system which admits a Lyapunov function which, by analogy with dissipative physical systems, may be called "energy". The evolution law is such that the energy is a non-

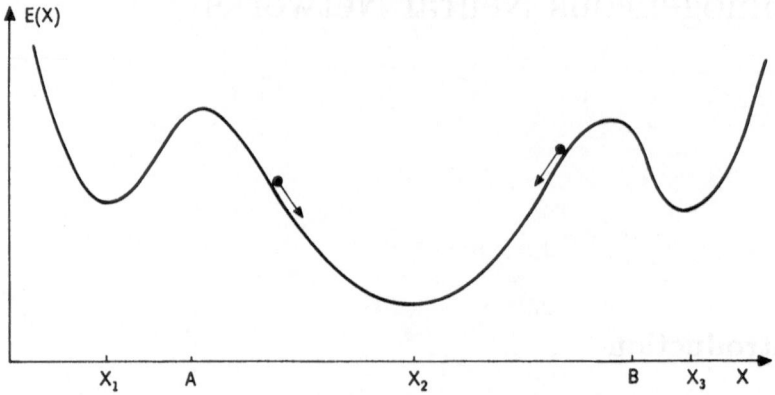

Fig. 4.1. The energy landscape. Nonlinear dynamical systems which admit a Lyapunov function are "initial condition recognizers". For example, each initial state $X \in]A, B[$ evolves towards the final state X_2. Note that the final state is not necessarily the nearest attractor in euclidean metric. By associating the energy function to a cost function, dynamical systems can be applied also to optimization problems

increasing monotonic function: in other words, the system evolves towards a local energy minimum, corresponding to a locally stable stationary state (see Fig. 4.1).

Optimization problems are characterized by the presence of a cost function, and consist of searches for states of the system which minimize this function. By identifying the energy with the cost function we can then (given suitable conditions) utilize a Hopfield network to solve optimization problems.

Hopfield and Tank (1985 and 1986) showed that, in many cases, it is possible to synthesize a neural network starting from the definition of an optimization problem. In this perspective, the relaxation of the network towards an attractor corresponds to the search for a minimum of the cost function. According to Hopfield and Tank, applications of the Hopfield model to optimization problems highlight the ability of the network to rapidly find sub-optimal solutions, even if it does not find the best solution. For many practical purposes, however, a rapidly found acceptable solution is much more useful than the optimum solution, if the latter requires extremely long search times. A discussion of the relative merits of the various optimization techniques, however, does not lie within the scope of the present work.

A further advantage related to the application of the Hopfield model – and of neural networks in general – to optimization problems derives from the possibility of their realization in electronic hardware or with optical devices (Denker, 1986) leading to enormous gains in computing speed with respect to computer simulation. The possibility of building the corresponding hardware is, in fact, another reason for the renewed interest in neural networks.

The next section of this chapter introduces the original Hopfield model, with binary variables, and discusses the dynamics and the memorizing and recognition capabilities of the model. This will be followed by a brief discussion of different possible variants of the original model with somewhat different cognitive properties. The two Appendices are devoted to a generalization of the original deterministic dynamics to a stochastic one (A4.1) and to the demonstration of some results quoted in the text (A4.2).

4.2 The Hopfield Model

In its original version, the Hopfield model is composed of a network of N boolean neurons, i.e., of binary elements which can take the values $X_i \in \{0, 1\}$, $i = 1, \ldots, N$.

Each neuron is connected to all the others through coupling coefficients (synapses). Let W_{ik} represent the synapse which carries the value of the k-th to the i-th neuron. The evolution rule of the system is extremely simple: at each time step, each neuron calculates the sum of its own inputs, $G_i(X)$, comparing this sum with its own threshold θ_i. Its value at the next time step will be 1 if the excitation exceeds the threshold, while it will be 0 if the excitation is below the threshold. In the limiting case where G_i equals θ_i the value of the neuron excitation remains unchanged. In formulae:

$$
\begin{aligned}
X_i(t+1) &= 1 && \text{if } G_i(X(t)) > \theta_i \\
X_i(t+1) &= 0 && \text{if } G_i(X(t)) < \theta_i \\
X_i(t+1) &= X_i(t) && \text{if } G_i(X(t)) = \theta_i
\end{aligned}
\tag{4.1}
$$

where the net input to the i-th neuron, also called its "local field", is given by:

$$
G_i(X) \equiv \sum_{\substack{j=1 \\ j \neq i}}^{N} W_{ij} X_j \ .
\tag{4.2}
$$

There are various versions of the model, which differ by the choice of behaviour in the case where the level of excitation equals the threshold (a case where the lowest output value is often chosen). In this section we will adopt the choice described by Eq. (4.1), which corresponds to the adoption of the so-called "majority functions", after the terminology of Fogelman (1986).

Different updating strategies can be chosen, depending upon the number of cells updated per unit time (Fogelman, 1986). If we choose a parallel iteration scheme, all the elements are updated at each cycle according to Eqs. (4.1) and (4.2), while at the other extreme (asynchronous updating) only one cell is updated at each time step. In this second case, the updating sequence may either be pre-defined or chosen randomly. Moreover, block-sequential updating strategies are also possible. In this section only random asynchronous updating will be considered. However, many results are similar to those of the other iteration schemes. It will be assumed here that the updating strategy "leaves no holes", i.e., that every cell is updated in a finite time (for finite N).

The asynchronous updating strategy, as we shall see, ensures a particularly simple dynamics, with convergence towards fixed points. This method is also widely adopted in the simulation of magnetic materials (Glauber, 1963). Simple models of the latter (Ising models) involve values of spin (permanent magnetic moments), associated with the various positions in the crystal lattice, which can only take binary values. In fact, in quantum mechanics, the measured spin of an electron, in an external magnetic field, can only be aligned with the field or with the opposite direction.

Spins can interact in different ways depending upon the material. The simplest situation is ferromagnetism, where the spins tend to align themselves, while in anti-ferromagnetic materials the interactions favour antiparallel neighbouring spins. Some materials, like e.g., spin glasses (Fischer, 1983; Mezard et al., 1987) show both types of interactions.

From these initial observations one can already see that both boolean neural networks and systems with spins of $\pm\frac{1}{2}$ may be described by binary variables, and that in both cases there are cooperative and competitive interactions. The similarity between the two cases is further illustrated in Appendix 4.1, where the stochastic generalization of the Hopfield model is described.

In fact, the deterministic dynamics described by Eqs. (4.1) and (4.2) may be considered as a limiting case of a more general stochastic dynamics. Taking the physical analogy of magnetic materials, the deterministic limiting case corresponds to considering the absolute temperature as tending towards zero, ignoring quantum fluctuations.

The correct generalization for including the case of finite temperature is discussed in Appendix 4.1. The concepts therein will be further examined in the discussion of the Boltzmann machine (Sect. 5.5).

It can be shown that, in the case where the matrix of the synapses W is symmetric, the system admits a Lyapunov function, i.e., a function which is bounded from below and remains non-increasing throughout the evolution of the system (we recall that we are considering here asynchronous updating).

This function (which has the same form as the energy of a spin system) is given by:

$$E(X) = -\frac{1}{2}\sum_{\substack{i,j=1 \\ j\neq i}}^{N} W_{ij}X_iX_j + \sum_{i=1}^{N}\theta_iX_i \quad . \tag{4.3}$$

$E(X)$ is non-increasing under the asynchronous dynamics defined by Eqs. (4.1) and (4.2):

$$E(X(t+1)) \leq E(X(t)) \quad . \tag{4.4}$$

Moreover, since $X(t+1)$ and $X(t)$ differ by at most one element:

$$E(X(t+1)) = E(X(t)) \Rightarrow X(t+1) = X(t) \quad . \tag{4.5}$$

In the model with asynchronous updating, the only attractors are, therefore, fixed points.

The demonstration of property (4.4) is straightforward. Let us suppose that the k-th neuron is chosen for updating at time t, and let $\delta X_k(t)$ be its variation, while $\delta E(t)$ denotes the corresponding variation of the energy function. Then:

$$\delta E(t) = -\frac{1}{2} \sum_{\substack{i,j \\ j \neq i}} W_{ij}[X_i(t+1)X_j(t+1) - X_i(t)X_j(t)] \cdot (\delta_{ik} + \delta_{jk})$$

$$+ \sum_i \theta_i(X_i(t+1) - X_i(t))\delta_{ik}$$

$$= -\delta X_k(G_k(X(t)) - \theta_k) \ .$$

If $\delta X_k(t) \neq 0$, it always has the same sign as $G_k(X(t)) - \theta_k$, therefore $\delta E(t) \leq 0$.

The Hopfield model is thus an example of the class of dynamical systems introduced in the previous chapters as candidates for carrying out cognitive tasks. It is a system with many degrees of freedom, which is nonlinear due to the presence of the threshold decision function. As we have also seen, its dynamics allows it to relax only towards fixed points. In order to use a system of this type for cognitive tasks, it must possess a sufficient number of attractors, and it must also be able to establish correct associations between initial conditions and final states.

In other words, the phase space portrait must be shaped in accordance with the cognitive task to be performed. Let us suppose (as in the example briefly outlined in Chap. 2) that the task consists in recognizing alphanumeric characters, even when they are distorted or incomplete. It must be stressed, however, that this is only an example with convenient visual features, these models being in no way limited to picture processing. In this type of example, the input configuration may consist of an imperfect letter (see Fig. 4.2) which is given to the Hopfield network as an initial condition. The system should then be able to relax towards an attractor representing the pure form of the letter, thus performing the required recognition.

Note that the two-dimensional representation of the state of the system in Fig. 4.2 is only intended to provide a visually interpretable pattern. Since each element in a Hopfield network is connected to all the other elements, there is no intrinsic topology, and the same result would have been obtained if, for example, all the elements had been placed along a line or in a three-dimensional cubic arrangement.

Let us suppose that we want to teach the system M different patterns $A^1, \ldots, A^M$, each one being an N-dimensional vector. The learning rule proposed by Hopfield is the following:

$$W_{ij} = (1 - \delta_{ij}) \sum_{m=1}^{M} (2A_i^m - 1)(2A_j^m - 1) \tag{4.6}$$

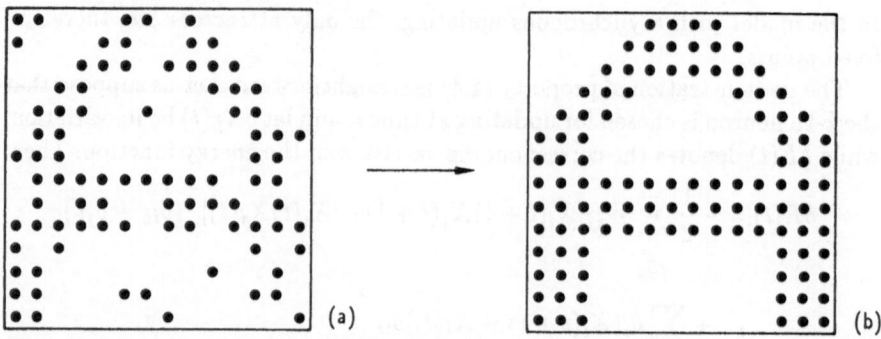

Fig. 4.2a,b. Dynamical evolution may correspond to recognition. In this case, an initial state (a), which represents a noisy picture of the letter "A", evolves towards a final state (b), which corresponds to a correct picture of that letter

where, as before, δ_{ij} denotes the usual Kronecker symbol: $\delta_{ij} = 1$ if $i = j$ and $\delta_{ij} = 0$ otherwise.

The intuitive idea behind this rule is related to the Hebb hypothesis introduced in Chap. 2. However, Hebb's original idea, which is limited to excitatory synapses, is formulated here in a rather particular manner. In fact, the Hopfield formula states that the connection between elements which are in the same state, within a given pattern, are reinforced, while those between elements which are in opposing states are lowered. Moreover, self-coupling is prohibited, i.e. $W_{ii} = 0$.

The reason why the Hopfield rule can reconstruct a noise-corrupted or incomplete pattern can be understood intuitively. For simplicity, let us assume that the system has been taught only one pattern which we will represent as a string of elements. Let the pattern be, for example:

$$* \ * \ * \cdots * \ * \ *$$

where "*" represents a value of 1 and "." a value of 0.

Let us take the case where all the thresholds θ_i are null, and assume that the input provided to the Hopfield network is an initial state which is an incomplete version of the learnt pattern:

$$* \ * \ * \cdots \cdots$$

If, at the first discrete time step, one of the first three elements is updated, it receives a net positive input due to the presence of the other two, whereas obviously elements with a value of zero do not contribute to the inputs of the other neurons: it thus remains in an ON condition. If one of the four central elements is updated, it receives an overall negative input (negative connections with the first three) and remains OFF.

If, however, one of the last three elements is updated, its overall input becomes positive, due to the positive connections with the first three elements, and will thus be switched ON. By applying this reasoning to the successive steps, it can be seen that the final state will be precisely the pattern already learnt.

Analogous considerations show that any element switched ON by noise, in a pattern similar to that already learnt, will be switched OFF. For example, a pattern such as

$$* * * \cdot \cdot * \cdot * * \cdot$$

will evolve towards the learnt pattern.

After having understood the intuitive reasons behind the learning rule (Eq. (4.6)), the effective capability of the Hopfield network to discriminate between different patterns has to be assessed. The key problems are therefore: i) to identify under which conditions the states $A^1, \ldots, A^M$ are fixed points of the Hopfield map, and ii) to determine their corresponding basins of attraction. To avoid ambiguities, the patterns $A^1, \ldots, A^M$, which appear in the synapses equations (Eq. (4.6)) will be defined from here onwards as "taught" or "stored" patterns. The term "learnt" patterns will only be used if they are fixed points.

Let us first define some useful quantities. The (unnormalized) overlap $Q(X, Y)$ between states X and Y is given by :

$$Q(X, Y) \equiv \sum_{k=1}^{N} X_k Y_k \quad . \tag{4.7}$$

The overlap function $Q(X, Y)$ measures the number of places where both patterns present high values. The overlap function can also be considered as a dot product of the two vectors X and Y. We also define the complement of an arbitrary pattern X, X', as:

$$X_i' = 1 - X_i \quad i = 1, \ldots, N \quad . \tag{4.8}$$

It should be noted that the matrix W whose elements are given by Eq. (4.6) and therefore also the time evolution of an arbitrary vector X ruled by Eqs. (4.1) and (4.2) are invariant under the transformation $A^m \to A'^m$. This means that teaching a given pattern or its "negative image" is exactly the same (Fig. 4.3).

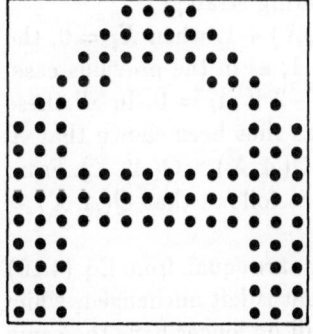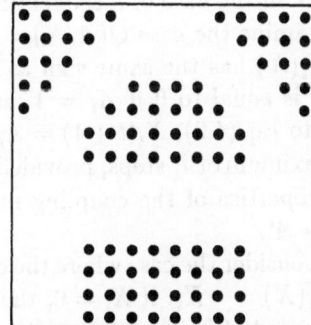

Fig. 4.3. A pattern and its complementary pattern

The synaptic matrix is also invariant with respect to the interchange of any two stored patterns. This allows us to extend to all the stored and complementary patterns the general results obtained for any one of them (i.e., those which do not depend upon specific hypotheses about that pattern).

The overlap function allows us to rewrite the input function $G_i(X)$, when the coupling matrix is given by Eq. (4.6), as :

$$G_i(X) = \sum_{m=1}^{M} (2A_i^m - 1)[Q(A^m, X) - Q(A'^m, X)] - MX_i \quad . \qquad (4.9)$$

In order to understand the behaviour of the Hopfield learning rule, let us consider the particularly simple case where only one pattern A has been taught; then

$$G_i(X) = (2A_i - 1)[Q(A, X) - Q(A', X)] - X_i \qquad (4.10)$$

Let us now examine the behaviour of this model in the case where all the thresholds θ_i are set to zero.

First of all it can be noticed that A is a fixed point, which can be proven by observing that $G_i(A)$ has the same sign as $2A_i - 1$, and therefore the evolution law leaves every element unchanged. From the previous remark about the invariance properties of the coupling matrix, it follows that A' is also a fixed point.

Another obvious fixed point is the null state $(0\ldots0)$. It will now be shown that they are the only fixed points: more precisely, an arbitrary initial state $X(0) = X$, different from the null state, evolves towards A if it overlaps with A more than with A', and it evolves towards A' if $Q(X, A) < Q(X, A')$. In the particular case $Q(X, A) = Q(A', X)$, the final state depends upon the first updated cell where X takes a nonzero value, say the k-th: if $X_k = A_k$, then $X \rightarrow A'$, otherwise $X \rightarrow A$.

In order to prove this result, let us first consider the case $Q(A, X) > Q(A', X) + 1$. In this case, $G_i(X)$ has always the same sign as $2A_i - 1$ and therefore $X_i(t+1) = A_i$. The final state A is reached in a number of iterations not greater than the number of steps R required to update every element at least once ($R = N$ in the case of a sequential updating strategy).

Let us now examine the case $Q(A, X) = Q(A', X) + 1$: when $X_i = 0$, the input function $G_i(X)$ has the same sign as $2A_i - 1$, as in the previous case; when $X_i = 1$, it is equal to 0 if $A_i = 1$, and to -2 if $A_i = 0$. In all these cases, according to Eq. (4.2), $X_i(t+1) = A_i$. It has thus been shown that X reaches A in a maximum of R steps, provided that $Q(A, X) > Q(A', X)$. From the invariance properties of the coupling matrix, it follows that $Q(A, X) < Q(A', X) \Rightarrow X \rightarrow A'$.

Finally, let us consider the case where the overlaps are equal: from Eq. (4.10) it follows that $G_i(X) = -X_i$. If $X_i = 0$, the element is left unchanged, while if $X_i = 1$ it is "turned off". Therefore, $X(t+1)$ will no longer have the same overlap with both A and A', since its overlap with the pattern which has a value of "1" in position "i" will be reduced, leaving the other unchanged. X will then evolve towards A, if $A_i' = 1$, and vice versa. This completes the proof of the statements made above.

If we had chosen a positive threshold, the consequence would have been that the null state would also have acquired a basin of attraction, consisting of all those initial states having similar overlaps with A and A'. It is also important to notice that convergence to the final state is very fast, requiring typically only one updating per site.

Let us now consider the case where two nonzero patterns, A and B, have been taught and $\theta = 0$. It can then be shown that A, B, A' and B' (which are called the "basic patterns") are always fixed points, together with the null state, which is, however, not attractive.

If a given initial state X overlaps with one of the basic patterns (say A) more than with any other, then $X \to A$ in no more than R steps.

If X has equal overlap with two basic patterns, the overlap with the others being less, then it will tend towards one or the other of these two patterns, depending upon the updating strategy. More precisely, the evolution is decided the first time a nonzero element (say, the i-th) is updated. If $X_i = 1$ and the two possible final states have different values, X evolves towards the final state which has a value "0" in that position. If both candidate final states have equal values in all those positions where X is nonzero, then the system evolves towards a final state which is the AND of the two candidates, i.e. it displays the value "1" only in those positions where the value of both states is "1". Thus, besides the basic patterns, the set of fixed points comprises all their AND's.

Finally, if X overlaps equally with all the basic patterns, then its evolution depends upon updating history. When for the first time an element with $X_i = 1$ is updated, it is always turned off, thus reducing the overlap with two non complementary basic patterns. From then on, the competition is limited to the other two basic patterns (as in the previous case).

The proof of these statements is lengthy, but can be made by generalizing the same techniques used in the one pattern case, as given in Appendix 4.2.

It must therefore be noted that, if only two different patterns are taught, they can always be made fixed points of the dynamics, by means of the Hopfield learning rule (Eq. (4.6)). However, in the two pattern case we have seen that the basic patterns (stored + complementary) are no longer the only fixed points, and so-called "mixed" states also appear, i.e. the AND's of the basic patterns.

In the case where the number of taught patterns is more than two, it is not possible to guarantee that they are fixed points. The reason is simple, and can be understood by considering the Hopfield learning rule, in the form given by Eq. (4.9). The input function is essentially a summation, over all patterns, of terms of the following type:

$$(2A_i - 1)[Q(A, X) - Q(A', X)] \ . \tag{4.11}$$

Take the case where the initial state X is equal to the memorized pattern A. The previous term then becomes:

$$(2A_i - 1)Q(A, A) \quad . \tag{4.12}$$

The effect of this term is to provide a positive (negative) contribution to all those neurons which have the value 1 (0) in pattern A, thus stabilizing the latter. The stationary properties of the basic patterns, in the cases studied above, are essentially based upon this effect.

The learning rule does not contain only one term, but a sum of similar terms (besides the term $-MX_i$). When there are two other terms (i.e., starting from the memorization of three patterns) there is no longer any guarantee that their combined effect will not dominate over the effect of the stabilizing term. Therefore, there is no longer the guarantee that the memorized patterns are fixed points.

However, some qualitative observations about the behaviour of the Hopfield rule in the case of more than one pattern can be made. Let us assume that the memorized patterns are generated in such a way that each of them (e.g., A) have about the same overlap with all the others (B) and their complementary patterns (B'). In this case the terms of the type $[Q(A, B) - Q(A, B')]$ are much smaller than $Q(A, A)$, and the stabilizing term dominates the summation. It is thus possible in this case to memorize various states. On the contrary, if two stored states are similar, they will give rise to significant interference effects, which destabilize both of them.

In order to quantify these properties, let us take a set of M random patterns. In particular, the elements of each pattern are generated in a completely random and mutually independent way from a uniform probability distribution:

$$p\{A_i^m = 1\} = p\{A_i^m = 0\} = \tfrac{1}{2}$$

for each $i = 1, \ldots, N$ and for each $m = 1, \ldots, M$. In this case, it can be shown that the maximum number of states which can be learnt ($M_{\max}$) increases with the dimensions of the network in a roughly linear manner (Hopfield, 1982):

$$M_{\max} \approx \alpha N$$

with $\alpha \approx 0.1$.

A more precise scaling law shows that in the limiting case where both N and $M_{\max}$ tend to infinity, the latter increases as $N/\log N$ (Weisbuch and Fogelman, 1985). Amit et al. (1985a and 1985b) have also provided an analytical treatment of the statistical mechanics of the Hopfield model, in the infinite N limit, describing its phase space properties as a function of a "temperature" (see Appendix A4.1). In particular, they determined the scaling parameter, showing that there exists a critical value for α (≈ 0.138) above which memorization properties are lost.

The conclusion is, therefore, that in the most favourable case the Hopfield network is capable of learning a number of patterns equal to about 10% of the number of its elements. However, an N-dimensional Hopfield network can code 2^N different patterns, and the computing load necessary for simulating the network on a serial machine increases as N^2. The poor storage capacity is therefore one of the limitations of the model. We have also seen that the system can not learn states which differ only by a few elements as independent attractors. Finally, it also seems rather unnatural to inevitably learn, together with every pattern, its complementary pattern, this also carrying the disadvantage of "filling" the phase space with attractors which have no particular meaning.

On the other hand, despite the above limitations, the Hopfield model is a simple and fairly "transparent" model of a neural network whose dynamics, as we have seen, can be well understood. The generation of spurious states, which causes serious problems from the applications point of view, could also be viewed as an expression of a kind of self-organization of the system.

Here we come across an example of the dichotomy, discussed in Chap. 1, between the "control" and "self-organization" points of view: in fact, the spontaneous generation of spurious states might also be regarded as an "interpretation" of the inputs autonomously developed by the system.

It should be noted, by the way, that this interpretation is not always "wrong" or "undesirable" according to the control point of view. If the training set contains a succession of "fuzzy" examples of the same pattern (e.g., different noisy versions of the same letter) the Hopfield model does, in fact, present the capacity to filter the noise, generating an attractor which coincides with the "clean" version of the pattern.

We can understand how this comes about by examining once again the form of the terms in the summation which appears in the local field at the i-th neuron. Let us suppose that the "clean" pattern is the vector U, and that the training set contains k fuzzy versions of U: $u^1,\ldots,u^k$. These are generated, starting from U, by altering the values of some of its elements chosen at random. To ensure significant cases, we will assume that the number of these modified elements is small, so that every u^m has a large overlap with U.

We will also assume that the other patterns in the training set do not exist, or that they do not significantly interfere with U ($Q(A,U) \approx Q(A',U)$ for each A). In this way it is sufficient to consider what happens to the k terms of interest. Suppose that the initial state is much more similar to U than to U', so that the overlaps with complements can be ignored. Since the various u^m are very similar to U, we can approximate $Q(u^m, X)$ with $Q(U, X)$, for each $m = 1,\ldots,k$. The input function will then be dominated by the expression:

$$Q(U, X) \sum_m (2u_i^m - 1)$$

From the assumptions made, and in particular from the low noise level hypothesis, the summation will be positive if $U_i = 1$, because in this case the positive terms will prevail, otherwise it will be negative (this is true for the limit $k \to \infty$, but in practice, for a modest noise level, a limited number of noisy versions is sufficient). In this way it can be seen that the state X is projected precisely upon the attractor U, "filtering out" the noise in the training set.

4.3   Modifications of the Hopfield Model

The original Hopfield model, as we have seen, gave rise to a great deal of interest in the scientific community and the severe limitations in its performance have stimulated further work, directed towards overcoming them.

Hopfield himself proposed one of the first modified versions of his own model, employing "analogic" neurons with continuously variable outputs instead of boolean ones (Hopfield, 1984; Hopfield and Tank, 1985 and 1986). The threshold firing rule (Eq. (4.1)) may be represented as a step transfer function (Fig. 4.4a), whereas the choice of analogic neurons allows the adoption of a continuous and differentiable S-shaped transfer function, such as that shown in Fig. 4.4b.

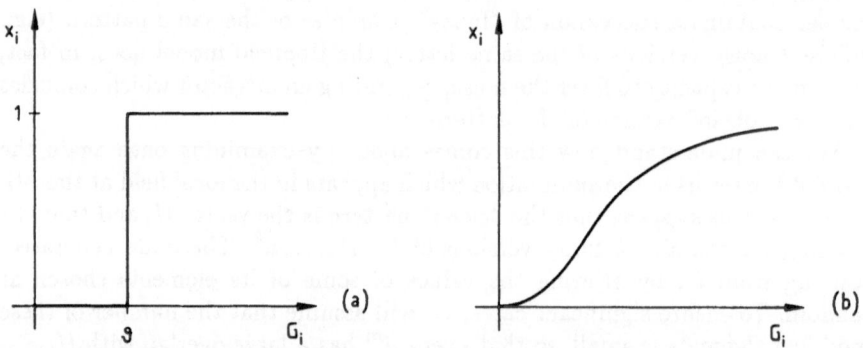

Fig. 4.4a,b. Threshold (a) and continuous (b) transfer functions

The precise form of the transfer function is not crucial, as long as it is S-shaped: for example, a logistic curve or a hyperbolic tangent may be used. The step transfer function may be considered as a limiting case of the continuous one, when the derivative becomes infinite.

Hopfield and Tank's model is "continuous" not only in the sense that it uses analogic neurons, but also because it employs a continuous representation of time, being based upon differential rather than difference equations. In the case where the connection matrix is symmetric, it can be shown that this network always evolves towards fixed points, since there is a Lyapunov function.

It was also possible to show (Hopfield, 1984) that the fixed points of the dynamics are often very close to those of the discrete model if the transfer functions are "steep" enough. The excitation values allowed in the continuous model lie between, say, 0 and 1, so that the phase space of the system is an N-dimensional hypercube of unit edge. The phase space of the boolean system, on the other hand, is represented by the set of vertices of this hypercube. This result thus implies that the energy minima of the Hopfield and Tank model are to be found, under the assumptions made, near the vertices of the hypercube.

The original reason for introducing the continuous model is both technological and biological: both biological neurons and amplifiers, in fact, have continuous transfer functions.

Also, the continuous model has turned out to be more useful for optimization problems. The Hopfield network relaxes towards local energy minima and in this way, as we have seen, it minimizes a cost function. Although the minima are very similar in both cases, Hopfield and Tank observed that it was easier, using the continuous model, to obtain more satisfactory solutions, i.e., corresponding to lower values of the cost function.

Another variation to the model was proposed by Dreyfus and co-workers (Personnaz et al., 1985 and 1986a). As we have seen, a crucial problem for the Hopfield model is the limited storage capacity: if the memorized patterns are more than two, there is no longer the certainty that these are fixed points of the dynamics.

However, by modifying the model, it is possible to guarantee that the taught patterns are fixed points, provided that one abandons the requirement of a local learning rule, that is, of a rule in which the value of the synapse between the i-th and the j-th neuron only depends upon the excitation values of those two neurons. If the main interest is directed towards artificial systems, this is not a problem of principle, but only a greater computing burden.

Let us take a boolean Hopfield model with values $\{ + 1, -1\}$ and zero threshold values: a pattern A is a fixed point if, and only if, for each $i = 1, \ldots, N$:

$$\left(\sum_{j=1}^{N} W_{ij}A_j \right) A_i \geq 0 \tag{4.13}$$

or, in vector notation:

$$WA = HA \tag{4.13'}$$

where W is the matrix of synaptic connections, A is the N-dimensional vector describing the learnt pattern, and H is an arbitrary diagonal matrix with non-negative elements. M arbitrary patterns will be fixed points if, and only if, a condition of the type in Eq. (4.13) is valid for each of them. By choosing, for simplicity, the arbitrary matrix H as the identity matrix, the condition for storing M patterns as fixed points is:

$$WK = K \tag{4.14}$$

where K is the matrix which has the patterns $A^1, \ldots, A^M$ as columns. We recall that our task is to find a learning rule, that is a synaptic matrix W satisfying the previous equation.

The solution can be obtained by using standard techniques of linear algebra (Kohonen, 1984), and is:

$$W = KK^I \tag{4.15}$$

where:

$$K^I = (K^T K)^{-1} K^T . \tag{4.16}$$

The superscripts I and T indicate, respectively, the Penrose pseudo-inverse matrix and the transpose. The pseudo-inverse matrix is a generalization of the usual notion of the inverse of a matrix, to which it reduces in the case where K is invertible, and which can be calculated by recursive methods (Kohonen, 1984).

In the case where the patterns A^m are mutually orthogonal, the matrix K is Hermitian and the prescription for learning becomes:

$$W = KK^T \tag{4.17}$$

which, in components terms, is:

$$W_{ij} = \sum_{r=1}^{M} K_{ir} K_{jr} = \sum_{r=1}^{M} A_i^r A_j^r \tag{4.18}$$

which coincides precisely with the Hopfield rule (in the formulation suitable for the choice of values $\{-1, +1\}$).

The rule proposed by Personnaz is thus a generalization of Hopfield's rule, and it is capable of guaranteeing the stability also for non-orthogonal states. This does not exclude the possibility, however, that other attractors besides the desired ones may be created. It should also be noted that the complements of the memorized states are, in their turn, fixed points.

The problem of spurious states was also examined by Kinzel (1985a and 1985b), who modified the learning rule by eliminating every frustrated connection (i.e., every connection which received both positive and negative contributions from different patterns). This measure, however, leads to a drastic reduction of the memorizing capacity of the network.

Amongst the mutually correlated patterns, particular attention should be given to the so-called "ultrametric" patterns (Rammal et al., 1986), which possess a natural hierarchical organization. A set of states in a normed space has an ultrametric structure if, taking any three states of the set (let these be a, b and c), two of the three distances between them are equal, that is (denoting the distance between x and y as $d(x, y)$):

$$d(a, b) = d(a, c) \quad \text{or} \quad d(a, b) = d(b, c) \quad \text{or} \quad d(a, c) = d(b, c) \quad . \tag{4.19a}$$

This condition is equivalent to the requirement that, for arbitrary states, the following relation holds:

$$d(a, b) \leq \max \{d(a, c), d(b, c)\} \quad . \tag{4.19b}$$

This can be demonstrated as follows: from the above definition of ultrametricity, Eq. (4.19b) trivially follows. In orded to prove that Eq. (4.19b) implies property (4.19a), let us assume, for example, that

$$d(b, c) \leq d(a, b) \quad . \tag{4.20}$$

Then

$$d(a, c) \leq \max \{d(a, b), d(b, c)\} = d(a, b) \quad . \tag{4.21}$$

But also:

$$d(a, b) \leq \max \{d(a, c), d(b, c)\} \quad . \tag{4.22}$$

On the other hand, $d(a, b)$ is less than neither $d(a, c)$ (Eq. (4.21)) nor $d(b, c)$ (Eq. (4.20)): thus the only possibility is that $d(a, b)$ is equal to $d(a, c)$ or to $d(b, c)$.

An ultrametric structure is a typical tree structure. By defining the distance between two leaves on a tree, at the same level, as the number of nodes which separate them from the closest common ancestor, it can be seen that each trio contains at least two elements which are equidistant from the third, thus forming an ultrametric structure (see Fig. 4.5).

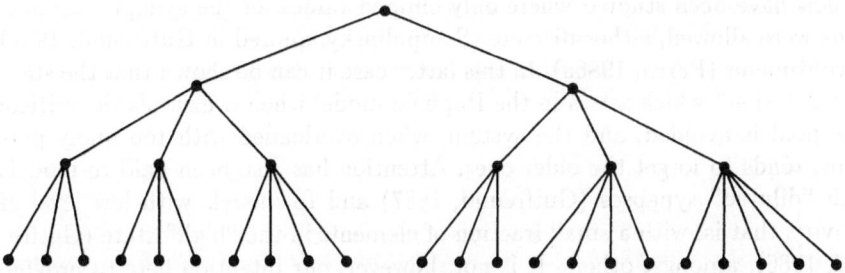

Fig. 4.5. The leafs of a tree form an ultrametric set

A hierarchical organization of the set of stored patterns corresponds to a structuring between very similar patterns (with high overlap) at a given level, less similar patterns at the next lowest level, etc. A simple way of generating ultrametric patterns (Parga and Virasoro, 1986) consists of starting from a given pattern, for example, a completely random sequence of $+1$ and -1, as the tree root. Let R_k be the value of the variable of the k-th cell. The next level is composed of a certain number of patterns, each of which is generated from the root by choosing, for each position, the value of X_k with the following probability distribution:

$$p\{X_k = 1\} = \tfrac{1}{2}(1 + \mu R_k)$$
$$p\{X_k = -1\} = \tfrac{1}{2}(1 - \mu R_k)$$

($\mu < 1$).

In this way the average (normalized) overlap between the root and its children is precisely equal to μ. In fact, the probability that two elements are the same is $\tfrac{1}{2}(1 + \mu)$, and that they are different is $\tfrac{1}{2}(1 - \mu)$. But the average overlap is precisely the expected value of the difference between the number of elements which are the same and those which are different: therefore, denoting expectation values with brackets $\langle \dots \rangle$, one obtains:

$$\langle Q(R, X) \rangle = \tfrac{1}{2}(1 + \mu) - \tfrac{1}{2}(1 - \mu) = \mu \quad .$$

The patterns of the first layer thus have an average (normalized) overlap with the root of $\mu < 1$. The patterns of the third layer are generated using the same procedure starting from the patterns of the second layer, and will thus have an overlap with the latter (their father) equal to μ, and an overlap μ^2 with the root. Gradually, the similarity to the root decreases on going along the levels.

Prescriptions for the memorizing of ultrametric patterns in Hopfield networks have been proposed by Parga and Virasoro (1986), Gutfreund (1986), Cortes et al. (1986).

Other variants of the Hopfield model have been proposed which take into account constraints of a biological and technological nature. For example, models have been studied where only clipped values for the synaptic connections were allowed, either discrete (Sompolinsky, quoted in Gutfreund, 1987) or continuous (Parisi, 1986a). In this latter case it can be shown that the state of "confusion" which arises in the Hopfield model when α exceeds the critical threshold is avoided, and the system, when overloaded with too many patterns, tends to forget the older ones. Attention has also been paid to models with "diluted" synapses (Gutfreund, 1987) and to models with low level of activity, that is, with a small fraction of elements in the "high" state (Gutfreund, 1986), amongst others. It is not, however, our intention here to provide an exhaustive review of all the variants of the Hopfield model which have been and which are still being proposed.

An important distinction has to be made, however, between models with symmetrical synaptic connections, which evolve towards fixed points, and models which violate this symmetry condition and in which more complex dynamical behaviours are therefore possible (Sompolinsky and Kanter, 1986; Parisi, 1986b; Sompolinsky et al., 1988).

A boolean network of N elements is a finite state automaton, which can be found in 2^N different states: its asymptotic states are therefore cycles of period not greater than 2^N or fixed points (i.e., "cycles" of period 1).

The fact that a Hopfield-like model with non-symmetrical synaptic connections can generate limit cycles can be easily understood. Take, for example, a matrix of synaptic connections which is quadratic in the components of the taught patterns, but which is not additive in the patterns themselves, containing interference terms. Let us assume that there are M different N-dimensional patterns $A^1, \ldots, A^M$, with components $\in \{0,1\}$, and that the matrix of the synaptic connections is:

$$W_{ij} = \sum_{m=1}^{M-1} A_i^m A_j^{m+1} + A_i^M A_j^1 \ . \tag{4.23}$$

Thus, the local field of the i-th neuron is:

$$G_i(X) = \sum_{m=1}^{M-1} A_i^m Q(A^{m+1}, X) + A_i^M Q(A^1, X) \ . \tag{4.24}$$

Let us now assume that all the stored patterns are orthogonal, or nearly orthogonal, so that their overlaps are small. Suppose then that the input state $X(0)$ has a large overlap with one of the stored patterns (say A^M) and an almost vanishing overlap with the others. Then the state at $t+1$ will be A^{M-1}, since the last term in the sum of Eq. (4.24) is much greater than the others. At time $t+2$ the state will be A^{M-2}, since only the last but one term is large, and so on. The form of the coupling rule (4.23) is such as to project every stored pattern upon another, thus giving rise to the sequence. A^1 is in turn projected to A^M, and this gives rise to a limit cycle behaviour. Obvious modifications of rule (4.23) can be considered, leading to shorter cycles.

It is interesting to notice that the set of stored patterns might give rise to competing cycles, and different input states would then lead to different cycles. Cycles may be interesting for information handling, because they can store more information than fixed points, for a given network size N (a cycle of length L totals NL bits, while a fixed point is limited to N).

A detailed discussion of attractor transitions, along with a wide reference list, can be found in Bell (1988).

We can now consider the problem of memorizing a number of patterns, which may be very similar, without formulating any restrictive hypotheses – for example, that they be organized in an ultrametric structure. A single Hopfield network can face this problem only by abandoning the locality of the learning rule, as in the model of Personnaz (Eq. (4.15)).

If, on the other hand, we consider local learning rules, in which the value of the synaptic coefficient W_{ik} is determined only by the values taken by the i-th and by the k-th element, we have to use a hierarchical structure such as that in Fig. 4.6, where the input state, to be recognized, is sent in parallel to several independent Hopfield networks ("subnetworks"). Each subnetwork will relax to the attractor closest to the input state, thus giving its own "interpretation" of the state. Competition amongst the different interpretations will be handled at a higher hierarchical level.

The simplest solution consists in presenting the various interpretations as input to the higher level, and in making a choice based upon some measure of similarity, such as the overlap. But, in many problems, it is possible to adopt more refined methods. In fact, one advantage of the hierarchical structure, compared to the single network, consists precisely in the possibility of allowing various alternative hypotheses to filter through to the next highest level. The choice may thus be made by combining an analysis of the input pattern with further information which may be available.

In the recognition of letters in a word, for example, the further information might be the recognition of the other letters, plus a dictionary. So, if the seventh letter in "connections" looks more similar to "*d*" than to "*t*", it will nonetheless be identified as "*t*" due to the contribution coming from the recognition of the other letters.

However, if we were to use the original Hopfield rule in the architecture of Fig. 4.6, we would be disadvantaged by the invariance of the synaptic matrix with respect to the transformation of any stored pattern into its complementary pattern: in fact, one cannot avoid teaching the complementary patterns. Therefore, the phase space becomes filled with useless patterns, which might sometimes be closer to a distorted input than to any "true" pattern. It would therefore be necessary to introduce a further check, which would be very expensive in terms of computational resources.

In order to avoid such an overloading of the higher level, and to avoid the rather inelegant check of the "admissibility" of the preferred attractor, one might try to modify the learning rule (Eq. (4.6)) in such a way as to maintain the attractive features of the original Hopfield rule while eliminating the complementary patterns.

For the time being, we will limit our attention to the consideration of symmetric synaptic matrices, thus ensuring that the only attractors are fixed points. The general form of a symmetric synaptic matrix, quadratic in the stored patterns, is

$$\Delta W_{ij} = \alpha A_i A_j + \beta(1 - A_i)(1 - A_j) + \gamma A_i(1 - A_j) + \gamma A_j(1 - A_i) \quad (4.25)$$

where ΔW is the variation of the coupling matrix when a new pattern A is taught. The original Hopfield rule corresponds to the choice $\alpha = \beta > 0$, $\gamma = -\alpha$.

However, the Hebb hypothesis does not compel us to choose exactly those values, if we treat the high and low states in differing manners. The Hebbian idea is to reinforce connections between neurons which "fire together", i.e. to take $\alpha > 0$. However, β is not necessarily constrained to take positive values, while $\gamma \leq 0$ (there is no point in reinforcing connections between neurons in different states). We can therefore identify six classes of generalized Hopfield rules: three possible choices for the sign of $\beta(+, 0, -)$ times two possible choices for $\gamma(0, -)$. However, the choice $\beta < \gamma$ clearly violates the Hebbian inspiration, thus leaving only five interesting classes.

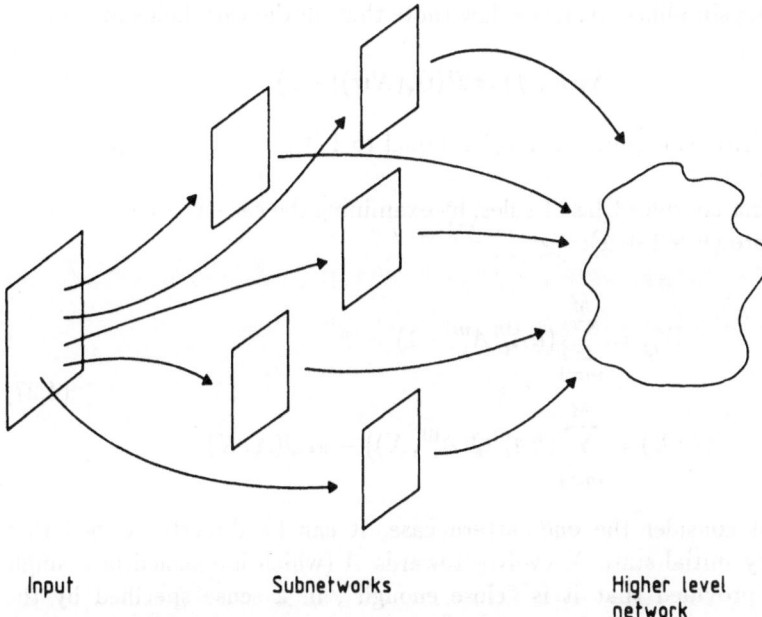

Input   Subnetworks   Higher level
                                                      network

Fig. 4.6. Hierarchical networks. The input is sent in parallel to several independent subnetworks ("Hopfield demons"). Each demon relaxes to its own fixed point, thus proposing its own "recognition hypothesis", which is sent to the higher level network for further processing

Since we are interested in breaking the symmetry with respect to the complementary patterns, we will require $\beta \neq \alpha$. Moreover, since we concentrate on the behaviour of classes of rules, we will impose the stronger requirement that $[\text{sign}\alpha] \neq [\text{sign}\beta]$, therefore $\beta \leq 0$. This leaves the following three classes:

Class 1: Only coincidences of high values give a positive contribution α, while both β and γ are negative.

Class 2: Only $\alpha > 0$, while the cases where at least one of the states is "off" do not contribute: $\beta = \gamma = 0$.

Class 3: Coincidences of high values give a positive contribution α, different values give a negative contribution $\gamma < 0$, while coincidences of low values have no effect: $\beta = 0$.

The third class, which has also been investigated by Toulouse et al. (1986), might actually be the closest one to the original Hebb idea.

For the sake of definiteness, here we will only consider parallel iterations and non negative thresholds, and we will not impose the condition that diagonal elements W_{ii} be set to zero: self-coupling is allowed. Moreover, we will consider the following simplified evolution law (note that all the thresholds are equal):

$$X_i(t+1) = H(G_i(X(t)) - \theta) \qquad (4.26)$$

where the Heaviside function $H(y)$ is equal to 1 if $y > 0$ and is equal to 0 if $y \leq 0$.

Let us first consider Class 1 rules, by examining the case where $\beta = \gamma = -1$ and therefore $(k \equiv 1 + \alpha)$:

$$W_{ij} = \sum_{m=1}^{M}(kA_i^m A_j^m - 1)$$

$$(4.27)$$

$$G_i(X) = \sum_{m=1}^{M}(kA_i^m Q(A^m, X)) - MQ(X, X) \quad .$$

Let us first consider the one-pattern case. It can be directly verified that an arbitrary initial state X evolves towards A (which is reached in a single iteration), provided that it is "close enough", in a sense specified by the threshold, i.e.

$$X \to A \quad \text{if} \quad kQ(A, X) - Q(X, X) > \theta$$

otherwise

$$X \to (0 \ldots 0) \quad .$$

The complementary pattern is therefore excluded from the set of attractors. Let us now consider what happens if two patterns A and B have been taught. By using techniques similar to those introduced in the study of the original Hopfield model, it can be shown that, on starting from an arbitrary initial state X, the next state can only be one of the following: $(0 \ldots 0)$, $(1 \ldots 1)$, A, B, U or Z, where U is the logical product (AND) and Z is the logical sum (OR) of A and B, i.e.

$$U_i = A_i B_i \qquad (4.28a)$$

$$Z_i = A_i + B_i - U_i \quad . \qquad (4.28b)$$

Depending upon the values of the parameters, different attractors may exist. A phase plot is shown in Fig. 4.7, as a function of θ and $Q(A, B)$, while Fig. 4.8 shows the final states which can be reached starting from some specified initial conditions, as θ and $Q(A, B)$ are varied.

It is interesting to note the following features, in the region where $\theta < Q(A, A)$, which is the interesting one:

i) at fixed θ, by increasing the overlap between A and B a transition occurs from a region where they are the only non-trivial fixed points to regions where other attractors appear (U, Z, limit cycles); finally, a region is reached

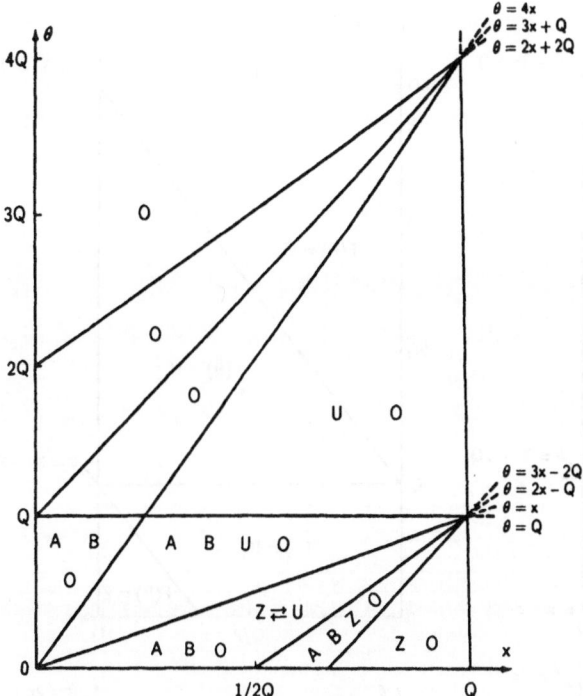

Fig. 4.7. Phase plot, Class I rule, two-pattern case. On the horizontal axis: the variable $x = Q(A, B)$. On the vertical axis: the threshold θ. The different possible final states are indicated within each region. The double arrow indicates a period-two limit cycle between states U and Z defined in Eq. (4.28). Here $Q(A, A) = Q(B, B) = Q$, and $k = 3$

where A and B are no longer independent attractors. In short, two patterns coexist as independent attractors if they are not too similar to each other.

ii) By increasing the threshold, with a fixed mutual overlap, the AND-type attractor is preferred over the OR-type.

Suppose now that "A" and "B" have a large mutual overlap, and that they have the same meaning: they might be, for example, slightly different drawings of the same alphabetical letter. In this case, they would give rise to mixed attractors, slightly different from both "A" and "B". The key word is "slightly": the subnetwork with "A" and "B" stored would relax to neither of them, but to mixed states very close to both. So, it would be able to recognize input states similar to the stored ones, even if its attractors were different from the original samples (recognition, in this case, is associated with relaxation to a non-zero state). In a sense, meaning is not lost, and the subnetwork will still perform precisely the task required by the architecture of Fig. 4.6.

Figure 4.9 shows an example of independent subnetworks trying to recognize a given input state; in every subnetwork only a single pure pattern is stored. It is shown how threshold modifications influence the selectivity

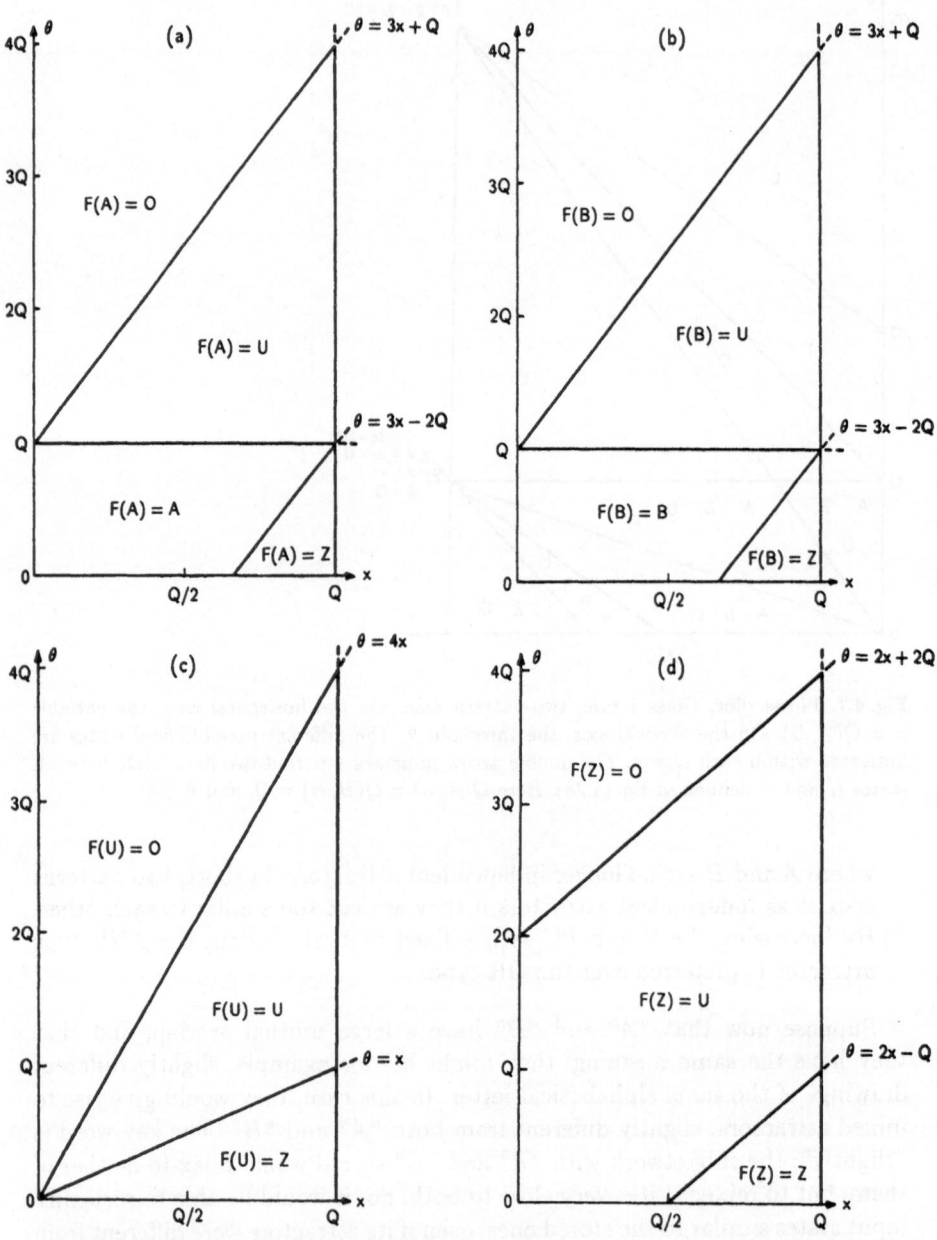

Fig. 4.8a–d. Class I rules, two-patterns case: final states reached, starting from some particular initial states, as a function of $x = Q(A, B)$ and of the treshold u. The initial states are A (a), B (b), U (c) and Z (d). $F(X)$ indicates the final state reached from the initial state X. The parameters are the same as in Fig. 4.7

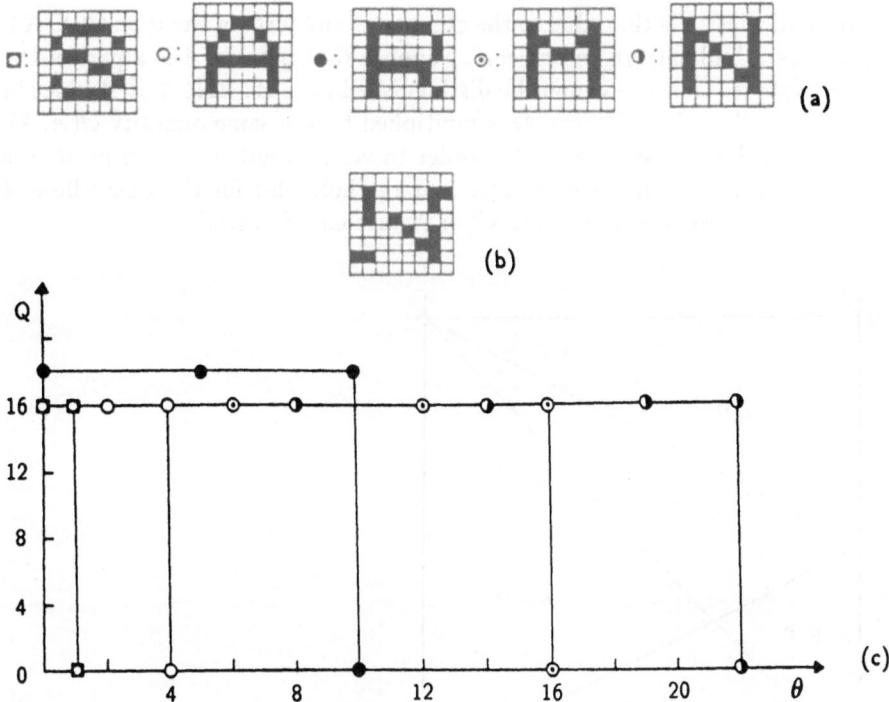

Fig. 4.9a-c. Competition among independent subnetworks, Class I rules. Each subnetwork stores only one pattern (a). The input state (b) is the initial state for every subnetwork. In (c), the auto-overlap $Q(Y, Y)$ of the final state Y is shown for every subnetwork, as a function of the threshold. Increasing the latter is a way to reduce the number of competing subnetworks (i.e., those which reach a final state different from the null state)

of the recognition process; indeed, Fig. 4.9 shows that threshold modification "controlled from above" can be an effective strategy to make subnetworks compete. It would be sufficient for the higher level subsystem to count the number of competing subnetworks (i.e., those relaxing to a non-zero state), and to increase (decrease) the threshold if the number is greater (less) than one.

The price which has to be paid for the elimination of complemetary patterns from rule 1 is the very limited storage capacity of random uncorrelated patterns ($M < k$). Class 2 rules can be analyzed using similar techniques. In this case:

$$W_{ij} = \sum_{m=1}^{M} A_i^m A_j^m$$

$$G_i(X) = \sum_{m=1}^{M} A_i^m Q(A^m, X) \quad . \tag{4.29}$$

Its phase plot, for the two-pattern case, is given in Fig. 4.10. The distinguishing feature of Class 2 is that, due to the absence of any term of the type $Q(X, X)$, it acts as a "subpattern recognizer". Suppose that pattern A is a part of the input state X; this case cannot be distinguished from the case $X = A$, since in both cases the "signal" term A_i is multiplied by the same quantity $Q(A, A)$. Therefore, this rule can be used in order to verify whether a given pattern is included in the input state. Similar remarks hold also for the case where A has a large overlap with a "feature" of X (see e.g. Fig. 4.11).

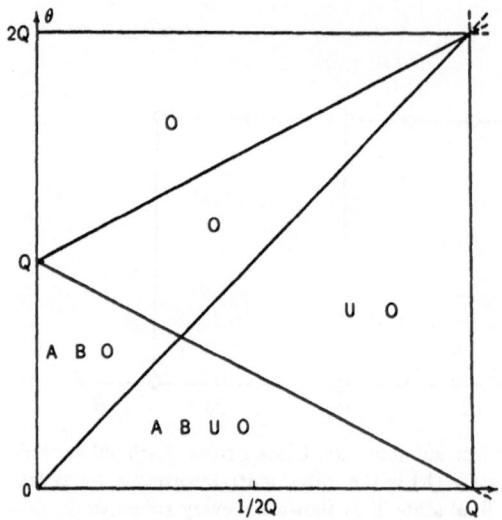

Fig. 4.10. Phase plot, Class II rules. Symbols are the same as in Fig. 4.7

Let us now turn to Class 3, whose equations can be written as :

$$W_{ij} = \sum_{m=1}^{M} (h A_i^m A_j^m - A_i^m - A_j^m)$$

$$(4.30)$$

$$G_i(X) = \sum_{m=1}^{M} \left[(h A_i^m - 1) Q(A^m, X) - A_i^m Q(X, X) \right] \quad .$$

In the one-pattern case, $X \rightarrow A$ if $(h-1)Q(A, X) - Q(X, X) > \theta$, otherwise $X \rightarrow (0 \ldots 0)$. In the two-pattern case a situation similar to that of Class 1 is encountered: U and Z are fixed ponts, A and B can coexist as independent attractors only if they are not too similar. However, a major difference is encountered, since other attractors are present, different from those described above, as can be seen from the phase plot of Fig. 4.12.

In order to clarify this point, we must precisely define the meaning of the idea of eliminating the complementary patterns from the phase space, besides the fairly obvious requirement that they should not be fixed points. In order to eliminate the complementary patterns, we must ensure that the final states are logical combinations of the patterns (like the AND and OR type of

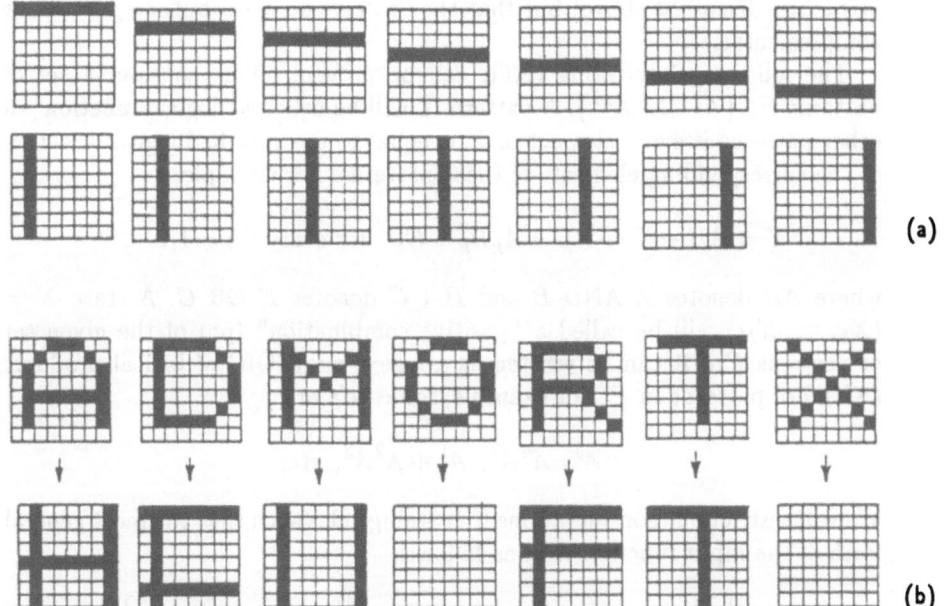

Fig. 4.11a,b. Class II rules are "subpatterns recognizers". In (a), the learnt states. In (b), initial states and the corresponding final states

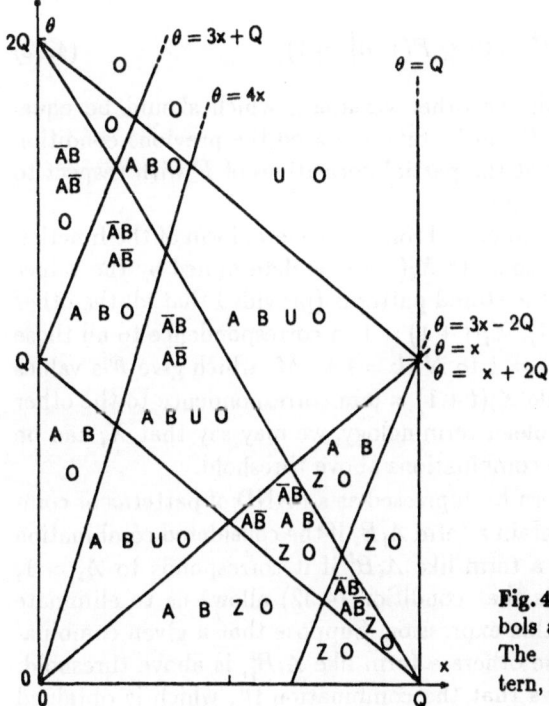

Fig. 4.12. Phase plot, Class III rules. Symbols are the same as in Fig. 4.7 ($h = 4$). The bar denotes a complementary pattern, and the product the logical AND

attractors discussed above) but that they receive no contributions from their complementaries.

The following interesting result can be proven: let us consider a set of patterns $P \equiv \{A^1, \ldots, A^M\}$. A pattern X will be called a "logical function" of other patterns if every element of X is equal to the specified logical function of the corresponding elements of those patterns. For example,

$$X = AB + C \Leftrightarrow X_k = A_k B_k + C_k \quad \text{for every} \quad k = 1, \ldots, N$$

where AB denotes A AND B and $B + C$ denotes B OR C. A state $X = (X_1, \ldots, X_N)$ will be called a "positive combination" (pc) of the given set of patterns P if it can be written as a logical sum (OR) of logical products (AND) of patterns in P. For example, some pc's are:

$$A^1, \ A^4 A^2, \ A^1 + A^2 A^3, \ \text{etc.}$$

We will restrict our considerations to learning rules such that the most general form of the input function X is as follows:

$$G_i(X) = F\left\{A_i^1, \ldots, A_i^M, Q(A^1, X), \ldots, Q(A^M, X), X_i, Q(X, X)\right\} \quad . \quad (4.31)$$

It can then be shown that, given an arbitrary state $X(t)$, its successor $X(t+1)$ is a pc of the learnt patterns $A^1, \ldots, A^M$, provided that, for every pair of indices i, k and for every X with components $\in \{0, 1\}$:

$$F(*, A_i^k = 0) \leq F(*, A_i^k = 1) \tag{4.32}$$

where the symbol $*$ means all the other variables, which should be equal on both sides of Eq. (4.32). If F can be differentiated the previous condition amounts to the requirement that the partial derivatives of F with respect to all the A_i^k's should be non-negative.

The proof can be outlined as follows. From the general form of the function F, it is seen that the value of the state $X_i(t+1)$ is determined by the values of the corresponding states in the stored patterns (provided that all the other variables in Eq. (4.31) are fixed). $X_i(t+1) = 1$ in correspondence to all those combinations of values of A_i^k, $i = 1$ to N, $k = 1$ to M, which give F's values higher than the threshold, while $X_i(t+1) = 0$ in correspondence to the other combinations. Resorting to boolean terminology, we may say that X_i can be expressed as the OR of all the combinations above threshold.

Every combination can in turn be expressed as an AND of patterns or complementary patterns: it will contain a term $A_i B_i$ if the considered combination corresponds to $A_i = B_i = 1$, a term like $A_i B_i'$ if it corresponds to $A_i = 1$, $B_i = 0$, etc. The thesis now is that condition (4.32) allows us to eliminate all the complementaries from this expression. Suppose that a given combination Ω, containing amongst the others a term like $A_i B_i'$, is above threshold; then, from Eq. (4.32), it follows that the combination Ω', which is obtained by changing B_i' into B_i, is also above threshold. The OR of these two combinations leads to the disappearance of the B_i value, which is not essential for overcoming the threshold, provided that the other terms are equal. In this way, all the terms containing complementary patterns can be eliminated.

Class 1 and 2 rules obey the condition (4.32), while Class 3 rules do not. Actually, descendants of complementary patterns are found in the set of attractors in this latter case (see Fig. 4.12).

Let us finally analyze the storage capacity of Class 3 rules. For this purpose we shall consider the case where the A's are random uncorrelated patterns: every element in every pattern is chosen at random independently from all the others. The probability distribution for each element is then assumed to be

$$p\{X_i = 1\} = f$$
$$p\{X_i = 0\} = 1 - f \ . \tag{4.33}$$

Therefore,

$$\langle Q(A^m, A^k) \rangle = f\delta_{mk} + f^2(1 - \delta_{mk}) \ . \tag{4.34}$$

In order to give a rough estimate of the storage capacity of this rule, one must determine the probability that a stored pattern is also a fixed point, i.e. (assuming $\theta = 0$ for definiteness) the probability that $G_i(A^k)$ has the same sign as A_i^k. From Eq. (4.29) it follows that

$$G_i(A^k) = ((h-1)A_i^k - 1)Q(A^k, A^k)$$
$$+ \sum_{m \neq k} \left\{ (hA_i^m - 1)Q(A^m, A^k) - A_i^m Q(A^k, A^k) \right\} \ . \tag{4.35}$$

The first term in the sum represents the signal term, since it always has the same sign as A_i^k, while the second term represents the noise, resulting from interference amongst the stored patterns.

An estimate of these terms can be given, by assuming that the stochastic variables A_i^m are independent from the overlaps $Q(A^m, A^k)$; in this case, Eqs. (4.34) and (4.35) lead to

$$G_i(A^k) \approx N((h-1)A_i^k - 1)f + N(M-1)(hf - 2)f^2 \ . \tag{4.36}$$

A similar analysis, performed on the original Hopfield model, leads to

$$G_i(A^k) \approx (2A_i^k - 1)Nf + N(M-1)f(2f-1)^2 \ . \tag{4.37}$$

While in both cases the signal term grows linearly with f, the structure of the noise term is different. They vanish for $f = 0$ (both), for $f = 1/2$ (Hopfield) and for $f = 2/h$ (Class 3); the last two values coincide if $h = 4$, which is our choice for this analysis. By studying the derivatives, it is easy to show that the Hopfield noise term is lower than that of Class 3 near $f = 1/2$ (its derivative vanishes), while the opposite happens near $f = 0$. This difference predicts a better storage capacity of the Hopfield model for $f = 1/2$, and the superiority of rule 3 for low f's; these predictions are confirmed by computer simulations.

We thus have a series of rules which show the desired absence of complements and their descendants in the set of attractors (Classes 1 and 2), but which have only modest memorizing capabilities. On the other hand, Class 3 rules show a reasonable memorizing capability for patterns with a low density of 1's but, although the complements of the stored patterns do not belong to the set of fixed points, some of their descendants do.

We will now present an asymmetric rule ($W_{ik} \neq W_{ki}$) which shows both of these characteristics: it is complement-free and has, at the same time, a similar memorizing capacity as Class 3 rules. It should be noted that the previous theorem, which provides the conditions for the elimination of the complementaries, holds unaltered even for asymmetric matrices: the demonstration, in fact, at no point uses the assumption that W is symmetrical.

The rule is as follows:

$$W_{ij} = \sum_{m=1}^{M} (1 - A_i^m) A_j^m$$

$$(4.38)$$

$$G_i(X) = \sum_{m=1}^{M} (1 - A_i^m) Q(A^m, X) \ .$$

The basic idea behind rule (4.38) is that, if neuron j is quiescent, no modification in the synapses from it to other neurons takes place: only active neurons influence learning, which takes place in the usual Hebbian style.

In the one-pattern case, A is the only non-trivial attractor, and its basin is composed of all initial states X such that $Q(X, A) > \theta$. The phase diagram for the two-pattern case is given in Fig. 4.13.

This rule is both complementary-free (because it satisfies the conditions of the previous theorem) and endowed with good storage capacity for the case where the patterns average value is small. By performing an analysis similar to that which led to Eq. (4.36) for Class 3 rules, one obtains the following estimate:

$$G_i(A^k) \approx N(2A_i^k - 1)f + N(M - 1)(2f - 1)f^2 \ .   (4.39)$$

The noise term vanishes at $f = 0$, $f = 1/2$. By comparing its derivatives with those of the Hopfield model (Eq. (4.37)) one finds that the latter grows faster at $t = 0$ and slower at $t = 1/2$, leading to a situation similar to that of rule 3. These crude theoretical estimates are at least in qualitative agreement with computer simulations, which confirm the superiority of rule (4.38) at low f's, and the superiority of the Hopfield model for $f \approx 1/2$.

Another interesting feature of this asymmetric rule is its capacity to handle the case of correlated patterns in an interesting way. We will restrict ourselves

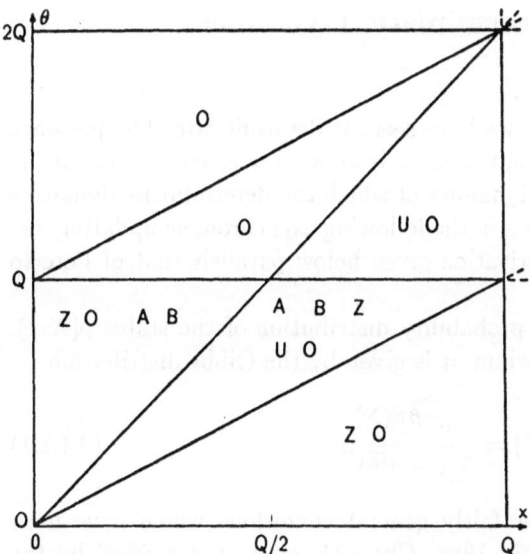

Fig. 4.13. Phase plot, asymmetric rule. Symbols as in Fig. 4.7

in the following to patterns with f, the fraction of pixels which are ON, smaller than $1/2$.

Suppose for instance that two patterns A and B have a finite overlap and are approximately orthogonal to all the other stored patterns, and let U be the pattern which is obtained by ANDing A and B. If the initial state has a much larger overlap with A or B than with any other pattern, we can limit our analysis to the subspace spanned by A and B, and the phase diagram of Fig. 4.13 applies. In particular, if $Q(X, B) - Q(A, X) > \theta$, $X \to B$, while if $\|Q(A, X) - Q(B, X)\| \le \theta$ and $Q(A, X) + Q(B, X) > \theta$, then $X \to U$. The system is capable of deciding whether a given input is closer to one of the stored patterns than to the other, or if it is impossible to decide between the two. In this case, it leads to the AND, which amounts to making a spontaneous classification (the input is in the same class as A and B, although it is neither of these). The interesting feature is that the new class U is generated by the system, without any need to be taught explicitly. This feature is already apparent in the Hopfield model, but less useful because of the presence of the complementaries.

Appendix 4.1 Non-Deterministic Dynamics of the Model

The original Hopfield model, as we have seen, is deterministic. The presence of the energy function is important because it allows the generalization of the model by defining a stochastic dynamics of which the deterministic dynamics is a limiting case. We will consider in the following asynchronous updating, i.e. one spin per time step. The derivation given below parallels that of Peretto (1984).

Let us try to determine the probability distribution of the states $p(X, t)$. In an isolated system at equilibrium, it is given by the Gibbs distribution

$$p(X) = \frac{e^{-\beta E(X)}}{\sum_Y e^{-\beta E(Y)}} \qquad (A4.1.1)$$

The evolution of $p(X, t)$, under fairly general conditions which have been discussed elsewhere (Serra et al., 1986, Chap. 1), may be described by the so called master equation:

$$
\begin{aligned}
p(X, t + \Delta t) &= \sum_Y w(X, \Delta t | Y, 0) p(Y, t) \\
&= w(X, \Delta t | X, 0) p(X, t) + \sum_{Y \neq X} w(X, \Delta t | Y, 0) p(Y, t) \quad .
\end{aligned}
$$
$$(A4.1.2)$$

From the normalization condition for the transition probability densities:

$$\sum_Z w(Z, \Delta t | X, 0) = 1 \qquad (A4.1.3)$$

we have

$$w(X, \Delta t | X, 0) = 1 - \sum_{Z \neq X} w(Z, \Delta t | X, 0) \quad . \qquad (A4.1.4)$$

By substituting in the master equation we obtain:

$$
\begin{aligned}
p(X, t + \Delta t) - p(X, t) &= \sum_{Y \neq X} w(Z, \Delta t | Y, 0) p(Y, t) \\
&\quad - \sum_{Z \neq X} w(Z, \Delta t | X, 0) p(X, t) \quad .
\end{aligned}
$$
$$(A4.1.5)$$

We then examine the conditions under which the Gibbs distribution is a stationary solution of the master equation:

$$0 = \sum_{Y \neq X} \left\{ w(X, \Delta t | Y, 0) e^{-\beta E(Y)} - w(Y, \Delta t | X, 0) e^{-\beta E(X)} \right\} \quad . \qquad (A4.1.6)$$

A sufficient condition for this is that every addendum of the sum in Y is zero, i.e.,

$$w(X, \Delta t|Y, 0)e^{-\beta E(Y)} = w(Y, \Delta t|X, 0)e^{-\beta E(X)} \quad . \tag{A4.1.7}$$

This condition, known as detailed balance, is satisfied if

$$w(X, \Delta t|Y, 0) = A(X|Y)e^{-\beta E(X)}$$
$$w(Y, \Delta t|X, 0) = A(Y|X)e^{-\beta E(Y)} \tag{A4.1.8}$$

provided that:

$$A(X|Y) = A(Y|X) \quad . \tag{A4.1.9}$$

The normalization condition can then be rewritten as

$$\sum_X A(X|Y)e^{-\beta E(X)} = 1 \quad . \tag{A4.1.10}$$

If all the coefficients $A(X, Y)$ are taken as being equal to a constant A, we have

$$A = \frac{1}{\sum_X e^{-\beta E(X)}} = \frac{1}{Z} \tag{A4.1.11}$$

where Z is the partition function.

As an alternative, it is possible to put some $A(X, Y)$ equal to zero without violating the condition of detailed balance.

According to the choices made, various dynamics are obtained. For example, one may impose that X and Y do not differ from one another for the states of more than a given number of cells. In the asynchronous dynamics the given number is 1. The idea is that, if the time increments are sufficiently small, it is plausible that only one neuron per step is allowed to update. Thus, $A(X|Y)$ is not zero only if the states X and Y differ at the most by one element. In this case the transition probability from the current state, say X, to the state Y is

$$w(Y|X) = \frac{e^{-\beta E(Y)}}{e^{-\beta E(Y)} + e^{-\beta E(X)}}$$
$$= \frac{1}{1 + e^{-\beta(E(X)-E(Y))}} \tag{A4.1.12}$$

This formula is the starting point for computer simulations. If one takes the limit $\beta \to \infty$, which is equivalent to the zero temperature limit ($\beta = 1/T$) one finds that the transition probability $W(Y|X)$ vanishes if the state Y has greater energy than the initial state X, whereas it takes a value of 1 if $E(Y) < E(X)$, in agreement with the deterministic dynamics of Eqs. (4.1) and (4.2).

Appendix 4.2 Memorization and Recognition of Two States

Let A and B be the two taught states. From the symmetry properties of the synaptic matrix W, it follows that A' and B' are also taught states. The form of W is:

$$W_{ij} = (2A_i - 1)(2A_j - 1) + (2B_i - 1)(2B_j - 1)$$
$$W_{ii} = 0 \quad . \tag{A4.2.1}$$

First of all, it can be shown that A, A', B and B' are stationary states of the network: it is sufficient to show this for one of these states, say A.

$$G_i(A) = 2(A_i - B_i)Q(A, A) + 2(2B_i - 1)Q(A, B) - 2A_i \quad . \tag{A4.2.2}$$

The table of the possible values of $G_i(A)$ is:

$A_i(t)$	$B_i(t)$	G_i	$X_i(t+1)$
0	0	≤ 0	0
0	1	≤ 0	0
1	0	≥ 0	1
1	1	≥ 1	1

It can be immediately seen that, at time $t+1$, each cell value is equal to A_i. Let us now take an arbitrary initial state X

$$G_i(X) = (2A_i - 1)(Q(A, X) - Q(A', X))$$
$$+ (2B_i - 1)(Q(B, X) - Q(B', X)) - 2X_i \quad . \tag{A4.2.3}$$

If the state X has a greater overlap with A than with B, A' or B', then, whatever cell X_i is updated, it takes the value of A_i. In this way the overlap with A is further increased, and the state evolves until it is coincident with A.

If X has equal overlap with two states A and B, and is less similar to their complementaries A' and B', then several cases are possible.

A_i	B_i	$X_i(t)$	$G_i(X)$	$X_i(t+1)$
0	0	0	≤ 0	0 still $Q(A, X) = Q(B, X)$
0	0	1	≤ 0	0 still $Q(A, X) = Q(B, X)$
0	1	0	$= 0$	0 still $Q(A, X) = Q(B, X)$
0	1	1	< 0	0
1	0	0	$= 0$	0 still $Q(A, X) = Q(B, X)$
1	0	1	< 0	0
1	1	0	≥ 0	1 still $Q(A, X) = Q(B, X)$
1	1	1	≥ 0	1 still $Q(A, X) = Q(B, X)$

Thus, if there are cells such that $X_i = 1$, $A_i \neq B_i$, either $Q(B, X)$ or $Q(A, X)$ does not vary while the other decreases: the symmetry is broken and X evolves until it coincides either with A or with B.

If there are no such cells, there is no way of breaking the symmetry, and since, in any case, X_i takes the value $A_i B_i$, X evolves until it coincides with $U(A, B)$.

Clearly, the symmetry of W once again allows exchanges between states, giving the family of stationary states $U(A, B)$, $U(A', B)$, $U(A', B')$ and $U(A, B')$.

In the further case where $Q(A, X) = Q(A', X) = Q(B, X) = Q(B', X)$, when the first nonzero cell, say the i-th, is updated, it unavoidably becomes 0, thus breaking the equality of all the overlaps, and falling into the previous cases.

The null state is stationary but is not obtained from different initial conditions.

5. Network Structure and Network Learning

5.1 Introduction

As we have seen in the previous chapters, dynamical systems may show non-trivial cognitive behaviour. Consider, for example, the Hopfield networks, dealt with in Chap. 4, which respond to the initial conditions by evolving towards a fixed point: they may be considered as cognitive systems capable of learning, and their memory is coded in the weights of the interconnections between elements.

A Hopfield network thus executes a particular cognitive task, namely the recognition of configurations. At this point, it is worth widening the issue, in order to briefly discuss the various cognitive tasks which can be carried out by a dynamical system.

A first class of cognitive tasks is that which may be defined as "recognition", or also "discrimination" (Anderson, 1986). In this case, the system is required to evolve towards the same final state starting from inputs externally defined as equivalent, and to evolve towards different final states starting from inputs externally defined as non equivalent.

A Hopfield network, for instance, carries out a recognition on the basis of the overlap of the pattern to be recognized with a "model" pattern, i.e., it refers to their scalar product. In order for the system to carry out recognition operations, it is therefore necessary, as we have seen, to externally code the model pattern in the connections of the network through learning.

In the presence of a single learning episode, the model pattern learnt in this way acts as an exemplar model for the recognition, in the sense that the recognition itself is based upon a comparison with the model. On increasing the number of examples of a certain class which are shown, the system tends to pass from the comparison with examples to a comparison with the "prototype", intended as the ideal example of a class, even though, as such, the prototype itself is not included amongst the examples shown to the system. This depends upon the fact that the details of the single examples tend to cancel each other, whereas the common elements are reinforced, as we have seen in Chap. 4.

A cognitive behaviour of this type may be defined as "generalization" (Anderson, 1986). The patterns taught, in fact, may be considered as examples out of a class: from their knowledge it is possible to identify other members of the same class to which they belong. It may also be stated that, on the basis of a sufficiently high number of learning episodes, the system is capable of developing inductive processes: i.e., starting from particular inputs of a class, it is capable of "inducing" the prototype of the class itself.

However, this kind of cognitive behaviour excludes any task which is not based on pattern overlapping or on "similarities" between patterns.

Take for example the recognition of the parity of a string of bits, which will be dealt with below in this chapter. In that case, the criterion of overlap has no value: in fact, strings which exhibit the greatest similarity (which differ only by one bit) have a different parity.

For this reason, a second class of cognitive behaviours must be considered: it may be defined as "feature extraction". In this case, we are no longer dealing with the execution of a cognitive task on the basis of overlapping patterns, but rather with the classification of the inputs as a function of some of their characteristics.

This second class of cognitive behaviours is, with respect to the previous one, closer to genuine conceptualisation. As we shall see, in fact, a system capable of carrying out a specific cognitive task belonging to this class must necessarily contain (or develop) some kind of more abstract internal representation of the task itself.

As we have seen in the previous chapter, there are two different stages in the functioning of a Hopfield network as a cognitive system. In the learning stage, specific input configurations are proposed to be memorised in the connections, whereas in the computing stage the network responds to the input configurations with a dynamics which depends upon the weights of the connections defined in the learning stage.

This kind of learning by examples may be defined "without supervision" (or "without reinforcement", or even "without feedback"). In fact, it does not include any procedure which, by feeding back a response to the system (be it the complete true solution, or simply a "yes" or "no" answer), reinforces (or weakens) the connections between units as a function of the success (or of the lack of success) of the recognition.

In other words, the network thus lacks the possibility to receive feedback from the environment in response to its own cognitive operations. This feedback interaction with the environment, which is not possessed by Hopfield networks, characterises on the contrary all natural systems with cognitive capabilities.

Hopfield networks, in their simplest configuration, are homogeneous networks: they have no structure, except for that defined by the interconnections. In particular, there is no functional difference either between cells or between sub-networks, in the sense that the input is provided as an initial condition for the network: it evolves towards an output which coincides with the steady configuration reached by the network itself.

As shown in the previous chapter, however, homogeneous networks of this type show cognitive abilities which are limited to configuration recognition.In order to obtain more complex cognitive behaviours (in particular, to realise a feature extraction behaviour) which may be modified through a learning procedure, systems with a more differentiated structure deserve consideration.

In this chapter, we shall then study structured networks, i.e. networks organized in layers of elements.

First of all, two-layer networks will be studied in Sect. 5.2, and it will be seen that they are limited to rather simple cognitive tasks. In Sect. 5.3 we shall then introduce learning algorithms for multi-layer networks, which prove to be more powerful, while in Sect. 5.4 we shall discuss some examples.

So far, we have considered only deterministic dynamical systems. The study of the characteristics and of the cognitive capabilities of probabilistic networks will then deserve a section by itself (Sect. 5.5), with the introduction of a suitable learning algorithm and the discussion of some examples.

Finally, we must note that to this point we have always considered supervised learning, i.e. learning with the presence of a teacher who sends to the system a suitable feedback. However, the possibility of designing network architectures capable of spontaneous learning is obviously of great interest, both from the theoretical and from the applicative point of view. For this reason, the last section of the chapter (Sect. 5.6) will be devoted to networks with unsupervised learning capabilities.

It must be stressed that this chapter does not intend to give a complete review of the field: it rather introduces some exemplar networks and discusses their characteristics from the perspective taken in the present volume. More detailed descriptions of the models which are briefly discussed here should be found in the quoted references.

5.2 Layered Networks

Networks may be subdivided into separate subnetworks on the basis of functional considerations. In particular, the presence of an input subnetwork and an output one is a preliminary condition for establishing a structured relationship with the external world and, in particular, to allow teaching interventions oriented towards reinforcing specific associations between stimuli and cognitive responses.

Every learning algorithm oriented towards the construction of associations between stimuli and responses must necessarily assume that both a particular stimulus and the corresponding response are "remembered" by the system when the system receives the instructive feedback from the teacher. This observation suggests the opportunity of providing the external stimulus as a time-independent input vector instead of as an initial condition, so that it may be present simultaneously with the output. In this case, the input is a permanent external perturbation: from the study of dynamical systems evolving from an initial condition which is their only input, as in the case of Hopfield networks, we thus move on to the study of forced systems.

This change, which may seem marginal, carries with it a significant transformation of the dynamical systems approach to cognitive processes, as presented so far. If, in fact, the input subnetwork has the only function of receiving the input information, like some kind of retina, we can imagine to ignore the mutual connections between its elements, and consider only those between input and output elements. The same is also true for the connections amongst the output elements: they appear, in fact, to be less interesting than those between the input layer and the output one. Moreover, "back-tracking" connections (from the output elements back towards the input ones) would introduce a kind of "unsupervised feedback" difficult to deal with. For this reason, we will normally ignore "backward" connections in the following discussion, including them only within a probabilistic approach (and in particular in the study of the so-called "Boltzmann machine" in Sect. 5.5).

So far, we have arrived at a system with two layers mutually interconnected in a one-way manner. The dynamical behaviour, so important in Hopfield networks, becomes rather trivial in this system. The dynamics, in fact, is reduced to a simple one-shot computation by each of the output elements. We may say, in general terms, that there is some kind of complementarity between the variety of dynamical behaviours and the programmability of cognitive processes, or, in other words, between self-organization and controlled learning. As we shall see, however, even in the controlled learning approach one again meets, in a different setting, the importance of self-organization processes in cognitive systems oriented towards feature extraction. This self-organization, as it will be shown, arises from the dynamics of weights adjustment.

First of all, let us consider a two-layer input/output system, and discuss a possible procedure for learning by experience based upon the relationship between stimulus and response. Let us thus suppose that we subject the system to a certain number of inputs: associated with each of these, the system receives in output the correct response which the system itself would be expected to reach if the weights of the connections were correctly adjusted. The aim of the learning process is precisely that of modifying the weights in a suitable way.

As in the case of Hopfield networks, we have here a learning stage distinct from the recognition one; the difference from Hopfield networks, however, lies in the presence of a feedback information. In fact, the output is a response which can be reinforced if correct, or discouraged if wrong, by acting upon the connections between elements of the two layers.

Correspondingly, Hebb's rule must be replaced with a learning algorithm suited to the new structure of the network. For the aim of reaching a practical definition of a learning algorithm, the so-called Widrow-Hoff rule (Widrow and Hoff, 1960), let us assume that every learning episode occurs in two stages. In the first stage, the input layer is clamped to a particular input vector, while the output layer is not allowed to evolve freely, being clamped to the answer which the system must learn. The first operation of the learning algorithm then consists in increasing the weights of the connections between active elements in the two layers by a quantity δ (we consider for the time being boolean neurons).

In the second stage of learning, the input layer remains clamped to the input vector, while the output is allowed to freely evolve towards a steady state. At this point, the learning algorithm decreases (by the same amount as above) the connections between active elements in the two layers. The net result of applying this algorithm is to leave the "right" connections globally unchanged; on the contrary, it decreases the "wrong" connections between active elements.

The following symbols will be used: $W_{ji}^{(s)}$ = weight of connection between output element j and input element i after the s-th learning episode; $\delta^{(s)}$ = increment to W_{ji} by the s-th learning episode; $d_j^{(s)}$ = desired response (corresponding to the j-th output element) which the system has to learn during the s-th learning episode, with

$$d_j^{(s)} \in \{0,1\} \ ;$$

$x_i^{(s)}$ = state of the i-th input element at the s-th learning episode, with

$$x_i^{(s)} \in \{0,1\} \ ;$$

$y_j^{(s)}$ = spontaneous response of the j-th output element at the s-th learning episode, with

$$y_j^{(s)} \in \{0,1\} \ ;$$

In the following, unless otherwise stated, we shall consider networks characterised by a spontaneous response given by

$$y_j^{(s)} = 1\left[\sum_i W_{ji}^{(s-1)} x_i^{(s)}\right] \tag{5.1}$$

where the function $1[\sum_i W_{ji}^{(s-1)} x_i^{(s)}]$ is 1 if the argument is > 0 and is 0 if the argument is ≤ 0.

It should be noted that this definition does not compel the adoption of a zero threshold for all the neurons. In fact, for example, to assign to the j-th neuron a positive threshold equal to α it is sufficient to add an element x_α, always in the "on" state, connected to the neuron j by a coefficient

$$W_{j\alpha} = -\alpha \ .$$

The learning algorithm may be mathematically expressed as

$$W_{ji}^{(s)} = W_{ji}^{(s-1)} + \delta^{(s)}\left(d_j^{(s)} - y_j^{(s)}\right) x_i^{(s)} \ . \tag{5.2}$$

A sequence of learning episodes of this type, however, is not capable of assuring the execution of any kind of feature extraction behaviour, as was proven by Minsky and Papert (Minsky and Papert, 1988). The problem, therefore, is that of designing a structure for the system which is more apt to complex cognitive tasks. This is obtained by introducing an intermediate structure (made of one or more layers) between the input and output layers. These layers are often called "hidden layers".

The need for an internal structure interposed between input and output for the execution of cognitive tasks of a certain complexity may be clarified through the use of a simple example.

Let us consider the boolean function of two variables XOR (exclusive OR). It is immediately clear that such a function can not be computed by a network with only input and output layers.

Let us recall, to this aim, that the output neuron only "fires" if the sum of the products of the inputs times the weights of the connections exceeds a pre-defined threshold. If this threshold is exceeded for the input combination $(1,0)$ or $(0,1)$, it will certainly be exceeded also for the combination $(1,1)$, making the computation of the desired boolean function impossible.

It is thus necessary to introduce an intermediate layer, functioning as a feature extractor. In this paticularly simple case, if also direct connections between the input and output layers are allowed, the internal layer may consist of one element only.

A possible configuration is shown in Fig. 5.1. When both inputs are zero, the intermediate unit (which has a positive threshold) is off. Thus, a zero signal reaches the output and, as in this case the threshold is positive, the output is zero. If only one of the two inputs is 1, the intermediate unit remains off and the output unit is switched on by the direct connection between input and output. Finally, when both inputs are 1, the intermediate unit fires to 1 and inhibits the switching on of the output unit. It can therefore be seen that the

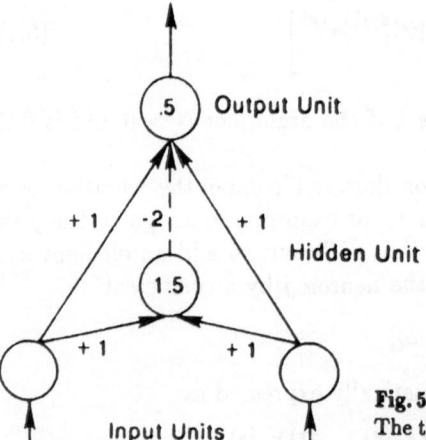

Fig. 5.1. An XOR network with one hidden unit. The thresholds are indicated within the small circles representing the units

intermediate unit performs the logical function A AND B on the two inputs A and B, while the whole system develops a particular realisation of the logical XOR function, i.e. XOR $(A, B) = (A$ OR $B)$ AND NOT $(A$ AND $B)$.

The logical function performed by the intermediate layer depends on the values and on the signs of the weights of the connections. Let us assume, e.g., the connections between input and intermediate unit and between intermediate unit and output as excitatory ones, and the direct input-output connections as inhibitory. In this case, if the thresholds are suitably defined, the intermediate unit carries out the logical OR function, the output then being clamped to zero by the joint inhibition of the two inputs when they are both equal to 1.

In a three-layer network, let us consider for a moment that both the connections between input layer and intermediate layer (feature extractor) and the threshold of the feature extractor are defined a priori. In this case, the state of the intermediate layer is determined by the input, without any kind of learning. Learning is therefore limited to a single set of weights: those connecting the intermediate and the output layer. According to the terminology introduced by Rosenblatt and extensively used by Minsky and Papert (Rosenblatt, 1962; Minsky and Papert, 1988), this is a "perceptron".

By means of a suitable combination of a priori design of the connections between input and intermediate layers and of learning through examples, it could be possible to teach to a suitably designed perceptron to compute any given logical function. Clearly, however, on increasing the complexity of the cognitive problem, the choice of the structure and of the connections of the intermediate layer (or layers) of such a network becomes increasingly critical. For this reason, the search for a learning by example technique with supervision without a complete pre-definition of the internal structure of the network deserves a particular interest. In the simple example just considered, this would involve, for the intermediate layer, the "spontaneous" emergence of the weights of the connections and of the threshold levels through the learning process, and thus the self-organized emergence of a particular internal representation.

5.3 Back-Propagation Algorithms

In the previous section we introduced a simple and intuitive algorithm for two-layer network learning (that is, networks with only one level of modifiable connections): the Widrow-Hoff rule. We may now look for a valid learning algorithm for multi-layer networks by trying to generalise this rule.

As above, let $x_i^{(s)}$ be the state of the i-th input element and $y_j^{(s)}$ be the state of the j-th output element in the s-th learning episode. For simplicity, let us consider a single layer of intermediate units, and let $z_k^{(s)}$ be the state of the k-th intermediate unit of this layer in the s-th learning episode.

The states of the units $y_j^{(s)}$ and $z_k^{(s)}$ are given by

$$y_j^{(s)} = 1 \left[\sum_k W_{jk}^{(s-1)} z_k^{(s)} \right] \tag{5.3}$$

$$z_k^{(s)} = 1 \left[\sum_i W_{ki}^{(s-1)} x_i^{(s)} \right] \tag{5.4}$$

where the function 1[.] is as defined above.

The Widrow-Hoff rule may be applied directly to the connections between the units of the intermediate layer and the output elements: in fact, the desired state $d_j^{(s)}$ for each output element is known. This is not possible, on the contrary, for the connections between input and intermediate layer. It should be noted once again that, in contrast to the case of pre-defined feature extractors ("perceptron-like" structures) the intermediate layer is here free to develop, through learning episodes, an a priori unassigned internal representation.

In order to arrive at a satisfactory defininition of a learning algorithm in multilayer networks, one must consider backward propagation of errors and not of desired states. The problem thus becomes that of the operational definition of an error, i.e. the choice of a "cost function" associated with the output error. Once this cost function has been defined, changes in the weights of the connections must be made in order to minimise its value.

The simplest possibile criterion is that of defining the cost of a particular response of the considered network to an assigned input

$$x^{(s)} = \{x_1^{(1)}, \dots, x_i^{(s)}, \dots\}$$

as the norm of the difference between the output vector

$$y^{(s)} = \{y_1^{(1)}, \dots, y_j^{(s)}, \dots\}$$

and the desired output vector

$$d^{(s)} = \{d_1^{(1)}, \dots, d_j^{(s)}, \dots\} \quad :$$

$$C^{(s)} = \|y^{(s)} - d^{(s)}\|^2 = \sum_j | y_j^{(s)} - d_j^{(s)} |^2 \quad . \tag{5.5}$$

The cost $C^{(s)}$ clearly depends upon the weights of the connections $W_{kj}^{(s-1)}$ and $W_{jk}^{(s-1)}$ as they are defined after the $(s-1)$-th learning episode.

Assuming that a sequence of m learning episodes is carried out, the weights of the connections should be defined through a learning procedure which, after the sequence has taken place, minimises the overall error

$$C = \sum_s C^{(s)} \quad .$$

(5.6)

However, an approximate error minimisation method has been proposed (Werbos, 1982; Rumelhart et al., 1986; Fogelman-Soulié et al., 1988) which introduces a correction to the the weights of the connections at each learning episode considered by itself. It is to be expected that an important role in the learning procedure will be played by the sequence of the learning episodes.

In order that the separation of the minimisation problem into sub-problems does not lead towards solutions too far from the optimum one, the single corrections are required to bring about a local cost reduction. A particularly simple and intuitive way to do this is to apply to each weight a correction proportional to the derivative of the cost $C^{(s)}$ with respect to the weight considered, i.e. to perform "gradient descent" in weight space:

$$W_{ij}^{(s)} = W_{ij}^{(s-1)} - \lambda^{(s)} \frac{\partial C^{(s)}}{\partial W_{ij}^{(s-1)}} \quad .$$

(5.7)

We now need to find an explicit expression for the derivative of the partial cost $C^{(s)}$ with respect to the weights. First of all, however, it must be observed that the function $C^{(s)}$ can not be differentiated, because it is made up of step functions 1[.]. To allow its differentiation, these functions must be substituted with smoother monotonic functions (e.g., sigmoids), which in the following will be generically represented by $f[.]$. Given that

$$A_i^{(s)} = \sum_j W_{ij}^{(s-1)} x_j^{(s)} , \quad x_i^{(s)} = f[A_i^{(s)}]$$

(5.8)

we can write

$$\frac{\partial C^{(s)}}{\partial W_{ij}^{(s-1)}} = \frac{\partial C^{(s)}}{\partial A_i^{(s)}} \frac{\partial A_i^{(s)}}{\partial W_{ij}^{(s-1)}} = \frac{\partial C^{(s)}}{\partial A_i^{(s)}} x_j^{(s)} \quad .$$

(5.9)

It can be shown that it is possible to find an explicit expression for the correction of the weights not only for the output layer, but also for the intermediate layer (within the simplified hypothesis of a single layer of this type). In order to do so, we must first compute

$$\frac{\partial C^{(s)}}{\partial A_j^{(s)}}$$

for the output layer cells, represented (as above) by the symbol $y_j^{(s)}$, while $d_j^{(s)}$ is the desired state for each one of them. Recalling the definition of $C^{(s)}$, one then has

$$\frac{\partial C^{(s)}}{\partial A_j^{(s)}} = 2 \mid y_j^{(s)} - d_j^{(s)} \mid \frac{\partial y_j^{(s)}}{\partial A_j^{(s)}} f'[A_j^{(s)}] \quad . \tag{5.10}$$

For the cells in the intermediate layer, represented by the symbol $z_k^{(s)}$, we have

$$\frac{\partial C^{(s)}}{\partial A_k^{(s)}} = \sum_j \frac{\partial C^{(s)}}{\partial A_j^{(s)}} \frac{\partial A_j^{(s)}}{\partial A_k^{(s)}} \tag{5.11}$$

$$\frac{\partial A_j^{(s)}}{\partial A_k^{(s)}} = \frac{\partial A_j^{(s)}}{\partial z_k^{(s)}} \frac{\partial z_k^{(s)}}{\partial A_k^{(s)}} = W_{jk}^{(s-1)} f'[A_k^{(s)}] \tag{5.12}$$

so that

$$\frac{\partial C^{(s)}}{\partial A_k^{(s)}} = f'[A_k^{(s)}] \sum_j W_{jk}^{(s-1)} \frac{\partial C^{(s)}}{\partial A_j^{(s)}} \tag{5.13}$$

It can thus be seen that a correction for the weights which depends on the gradient of the cost function is first computed for the output layer and is then propagated back to the intermediate layer. The case of two or more intermediate layers can be studied in an analogous manner.

The choice of the value to be assigned to $\lambda^{(s)}$ in Eq. (5.7) is conditioned by contrasting considerations.

It will be recalled that the procedure proposed for the back propagation algorithm does not correspond to the minimisation of an overall cost function, but to a reduction, on the basis of an evaluation of the gradient, of the contribution to the error carried by the individual learning episodes.

Updating the weights after the presentation of each pattern corresponds qualitatively to introducing a sort of noise in the gradient descent procedure for minimizing the total error function. Indeed, the single corrections tend to align themselves to the global one, but with superimposed fluctuations.

The convergence of this method, however, may be slow; moreover, there is the ever-present danger of being trapped in a local minimum. We will examine this problem in more detail in Sect. 5.5, devoted to the Boltzmann machine.

5.4 Self-organization and Feature Extraction

This section presents some results obtained by various authors through the application of the back propagation algorithm to three-layer networks.

We begin with the logical XOR function, already mentioned above. Fig. 5.2 shows some result obtained by Rumelhart et al. (Rumelhart et al., 1986).

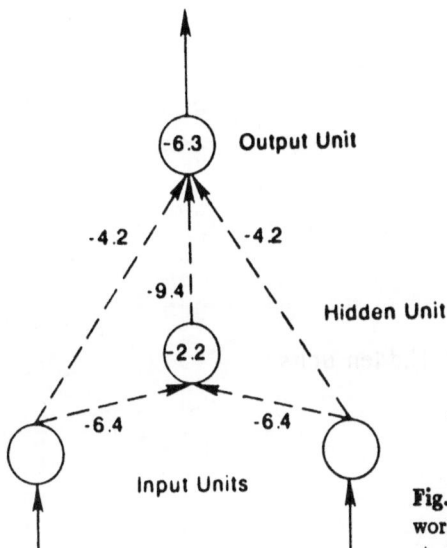

Fig. 5.2. Experimental results on an XOR network with one hidden unit (data from Rumelhart et al., 1986)

In this example, the XOR function is realized in a different manner with respect to the intuitively more simple cases described previously. The intermediate unit, in fact, performs a negated OR of the inputs instead of an AND or an OR. The system thus builds its own representation of the problem, orienting the intermediate unit towards the detection of a particular input feature.

The concept of internal representation becomes clearer when one considers an internal structure in which the intermediate layer is interposed between input and output, avoiding direct I/O connections (Fig. 5.3).

With reference to an architecture of this type, let us now consider the classical problem of determining the "parity" of strings of binary units. It may be useful to recall that parity is a binary variable which is zero or one according to wether the string contains an even or an odd number of ones. This problem is particularly suitable for illustrating the conceptual difference, within the field of cognitive activities, between recognition and feature extraction. In this case, the difficulty of the cognitive problem lies precisely in the fact that the patterns which show the greatest similarity (those which differ by only one bit) have a different parity.

The XOR problem discussed above may be considered as a particular case of a parity problem for an input pattern consisting of only two bits.

The problem of parity determination for strings of up to six bits has been thoroughly studied (Rumelhart et al., 1986). Figure 5.4 shows schematically a type of internal representation which, according to Rumelhart et al., is sometimes developed by the system in its intermediate layer for an input string of length N, in the case where also the intermediate layer is composed of N elements.

Output

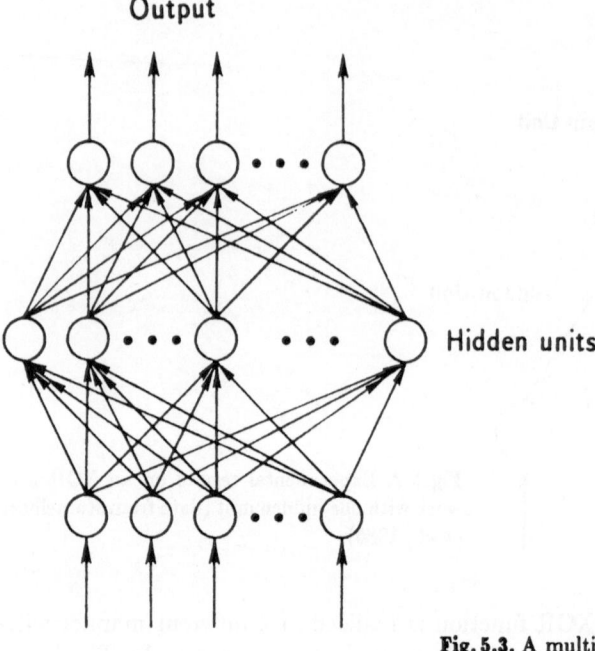

Hidden units

Fig. 5.3. A multilayer network without direct connections between input and output

Input

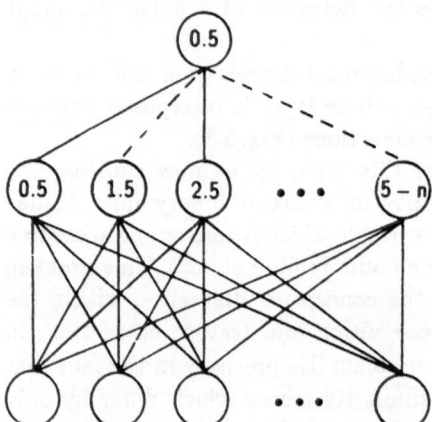

Fig. 5.4. A solution to the parity problem discovered by a trained learning systems (data from Rumelhart et al., 1986)

It can be seen that the units in the intermediate layer (the "hidden units") organize themselves in order to count the number of 1's. In particular, the first hidden unit on the left of the figure switches on in the presence of one or more 1's, the second in the presence of two or more 1's, and so on. Since the connections between the hidden units and the output units are alternatively positive and negative, the output is off for an even number of 1's in input, whereas it is on for an odd number.

Worth stressing is the fact that the internal representation which develops spontaneously in this case may be considered as an elementary form of conceptualisation; it does not, in fact, depend upon the position of the 1's and 0's in the input pattern, but only upon their overall number.

It should however be mentioned that the elegant solution of Fig. 5.4 is not always found by the system, which sometimes discovers more clumsy and hard-to-read representations.

Let us now consider another cognitive problem whose solution depends critically upon the development by the system of a feature extraction rather than of a pattern matching capability: the problem of recognition of the symmetry of a binary input string with respect to its centre. Here again, very similar patterns may require a completely different classification in terms of their symmetry.

It has been shown that two hidden units are sufficient to solve the symmetry problem (Rumelhart et al., 1986). The internal representation developed by the system may be understood with the help of Fig. 5.5, which refers to a six-bit input string. For greater clarity, the two units of the intermediate layer are shown at the top and at the bottom of the figure, while the output unit is on the extreme right.

It can be seen that every input unit is connected to the two units of the intermediate layer with equal weights but with opposite sign. Also, the weights corresponding to the units which are symmetrical with respect to the centre of the pattern are of equal value and of opposite sign. Thus, when a symmetrical

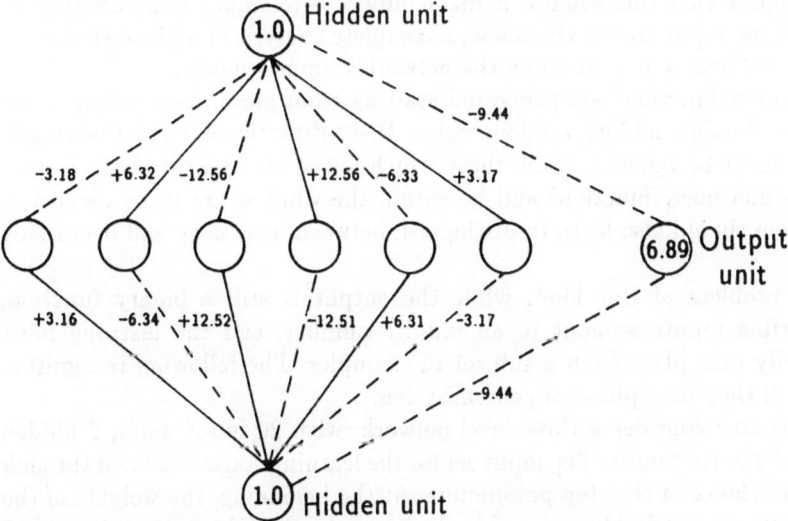

Fig. 5.5. A network for solving the symmetry problem. The six open circles represent the input units. There are two hidden units, one shown above and one below the input units. The output is shown to the right (data from Rumelhart et al., 1986)

pattern is presented to the system, both intermediate units receive a null (or approximately null) input and, as they have a positive threshold, they are both off. In this case, the output unit, which has a negative threshold, is on. It can then be observed that the weights corresponding to the various units do not all have the same value, but are approximately in the ratio of 1 : 2 : 4 in each semi-pattern; this allows a different activation signal for each of the possible configurations and, in particular, avoids random balancing between non-symmetrical semi-patterns. Finally, we note that the two hidden units receive the same activation signal, but with opposite sign: thus, when any non-symmetrical pattern is present, one of the two intermediate units is on, consequently turning off the output unit.

So far, we have focused our attention upon a particular class of cognitive tasks: the extraction of logical features. This has allowed a clearer illustration of the birth of internal representations. It is important, however, not to lose sight of the fact that multi-layer networks with a back propagation algorithm are capable of carrying out feature extraction tasks of a more general type, i.e., not limited to logical functions.

Let us now consider a problem of a different kind: the recognition of a step discontinuity in an input function (Malferrari et al., 1989). Let us assume that the input function is defined in a number of points equal to the number of input units (whose values are no longer restricted to 0 or 1). The signal lies within a given interval; moreover, a "minimum discontinuity" is defined, such that larger discontinuities are considered meaningful, while smaller ones are ignored. The network's task is defined as the identification of discontinuities which take place inside a pre-defined "window". In the following example, we shall suppose that this window is made only of three input units around the center of the input string. Of course, a complete analysis of a given signal can be achieved letting it shift along the network's input window.

The input functions are generated starting from piecewise constant functions and further adding random noise. Discontinuities outside the chosen window must be ignored, while those which are above the threshold (in the original, non noisy function) and lie within the window are to be identified; the system should also learn to distinguish between increasing and decreasing ones.

In a problem of this kind, while the output is still a binary function, the possible inputs amount to an infinite number, and the learning must necessarily take place from a sub-set of examples. The following recognition phase will then be a phase of generalisation.

Let us now consider a three-level network, with 20 input units, 2 hidden units and 2 output units. The input set for the learning stage is defined through a random choice of the step parameters. At the beginning, the weights of the connections are randomly assigned in the interval $[-1, 1]$. In the learning stage, a set of randomly chosen steps is presented; each time, the weights are updated following a back propagation algorithm (see Sect. 5.3).

After some learning cycles, the network reaches a stable configuration with a recognition rate of almost 100%. This configuration may be interpreted as corresponding to a different specialisation of the two hidden units.

More precisely, we see that one of the two hidden units specialises as a detector of differences in mean value between left and right units. For this units, the weights of the connections with the input layer will be as shown in Fig. 5.6a. The second hidden unit specialises as a discontinuity detector: since the discontinuity is allowed only around the central input unit, the weights of the connections between this hidden unit and the input layer will be as shown in Fig. 5.6b, or else with opposite signs.

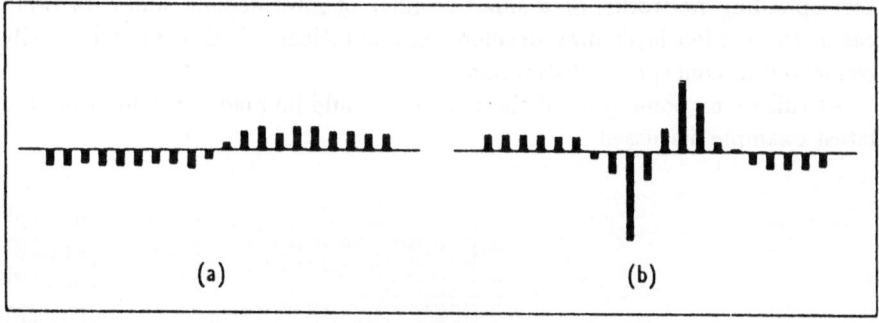

Fig. 5.6a,b. The weights of the connections between the input units and the two hidden units (see text) in an edge detector (after Malferrari et al., 1989). The network has three layers with 20, 2 and 2 elements. Two output neurons are required to code for three possibilities: increasing, decreasing or non-existing discontinuity.

The weights of the connections from the hidden units to the output units then adjust themselves to give the required answer (i.e., one or the other output unit on if there is a rising or falling discontinuity).

This specialisation of the hidden units corresponds to a symmetry breaking. Each of the two functions described will then be realised by one or the other of the hidden units according to the values of the randomly assigned initial weights. In some cases, however, the hidden units do specialise in a symmetrical mix of the two previously defined functions.

A simple and clear example of another task different from a purely logical one can be given by a speech recognition task (Elman and Zipser, 1987). In particular, let us consider the recognition of the consonants "b", "d" and "g" starting from examples of sounds co-articulated with vowels ("ba", "bi", "bu", "da", "di", "du", "ga", "gi", "gu"). Obviously, the co-articulation renders the consonant recognition more difficult.

Also in this case, the network consists of three interconnected layers. The input layer is a matrix of cells upon which a picture of the Fourier transform of the signal is constructed: i.e., a generic input cell contains the spectral intensity corresponding to one of the chosen frequency bands and to one of the given time intervals. The output consists of three cells, each of which codes the presence of one of the consonants. The intermediate layer also consists of three cells, connected to both the input and the output; through a back-propagation algorithm, they develop a distributed representation of the consonants, as shown in Fig. 5.7. The figure shows that the representation retains trace of the co-articulation, but succeeds in representing, in a clearly different way, the three consonants.

It should be observed that in all the examples considered it was rather easy to "translate" the internal representation developed by the intermediate units into a conceptual representation at a symbolic level. This task was surely facilitated by the choice of the problems and by the presence, in the corresponding networks, of a small number of intermediate units. In other cases, the hidden layer may develop representations which cannot be easily connected to conceptual abstractions.

At this point, some general observations should be made, starting from the latest example discussed.

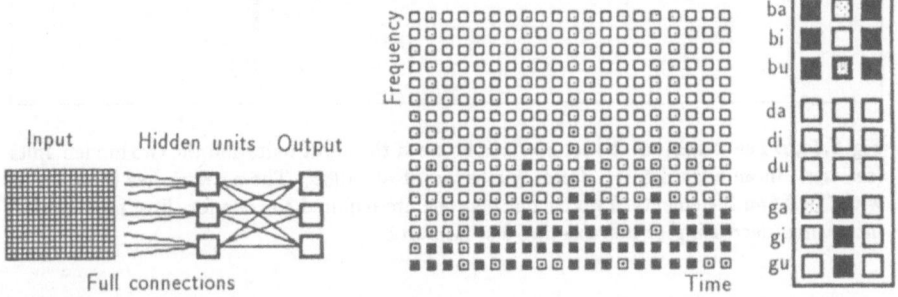

Fig. 5.7. A network for speech recognition. The figure shows a network with one hidden layer (left) which was trained to recognize three consonants from sound patterns, represented at the input level by 20 × 16 real values. Middle: a "ba" sound is shown. The blackness of the squares is proportional to the energy in the various frequency intervals (after Elman and Zipser, 1987)

In that case, the set of possible situations to be identified is potentially infinite, although subdivided into nine different classes. In situations of this type, clearly one cannot construct the learning process through a presentation of all possible inputs. It is therefore particularly important that the examples chosen for learning are representative of the inputs which will afterwards be presented to the network for recognition. The quality of learning may then depend upon the examples presented to the system during the learning stage.

In all the examples considered, the definition proposed for the cost function turned out to be acceptable. In other cases, with more complex outputs, the cost function might need to be re-defined, for example by weighting differently the discrepancies at the various output cells.

The question about the extent to which the results obtained so far can be extended to problems of larger size remains, at present, open. In order to make some comments on this point, we shall refer once again to the case of determining the parity of a string of bits. As we know, once suitably instructed, the system with a hidden layer may spontaneously develop the "conceptual architecture" shown in Fig. 5.4. Minsky and Papert (Minsky and Papert, 1988) note however that this result has the limitation of implying a fast growth of the number of weights; there are thus practical difficulties for its application to examples of greater dimension. A special purpose structure, such as that shown in Fig. 5.8, which may be simply obtained by using an intermediate layer with pre-defined connections, appears on the contrary to avoid these limitations. A similar comment can be made about the other examples studied. In other words, the back-propagation learning procedure does not guarantee that the system will orient itself towards efficient solutions which can be extended to large size problems; even more so if one takes into account the possibility of being trapped in a local minimum of the cost function.

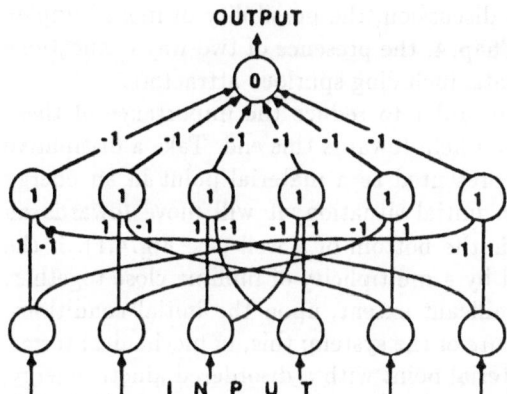

Fig. 5.8. A special purpose network for parity determination (after Minsky and Papert, 1988)

Minsky and Papert are thus quite cautious about the possibility of extending to large size problems the back-propagation learning approach presented above. They prefer the hypothesis that these problems should be approached by designing networks whose components are in turn relatively small networks, with a large range of architectures and operational modes.

This is a rather interesting and stimulating proposal. It does not, however, reduce the conceptual and heuristic validity of the research, presented in this chapter, on structured networks and on their learning processes. Moreover, it must be said that in a number of "real world" problems (particularly in the field of signal and speech processing) back-propagation networks can compete with the most widely adopted pattern recognition algorithms. A discussion on this point will be developed in Chap. 7.

5.5 Learning in Probabilistic Networks

It will be recalled that the hierarchical networks dealt with in the preceding sections were characterised, unlike Hopfield networks, by one-way connections from input to output. This choice was determined by the need to find a learning algorithm making use of the instructive nature of feedback.

In this way, the range of potential dynamical behaviours of the system was reduced to favour an auspicated "cognitive plasticity". However, it is interesting to investigate the possibility of a complementary choice in the direction of maintaining a greater variety of dynamical behaviours in structured networks.

Let us again consider networks with hidden units, since we already know that only very simple cognitive behaviours are possible in input-output networks. Now, however, all the connections will be assumed to be two-way and symmetrical.

Assume now that an input is sent to the system by clamping the input units. The system, through a pre-defined updating procedure (e.g., random updating of one unit at a time), will then relax towards a fixed point. For the sake of simplicity, we will ignore, in this discussion, the possibility of more complex behaviours. As we have seen in Chap. 4, the presence of two-way connections allows a wide variety of fixed points, including spurious attractors.

The system can be modified in order to reduce the importance of these problems: reasoning by analogy can help towards this end. Take a dissipative physical system which can be represented as a material point in an energy landscape. Starting from a given initial situation, it will move towards an energy minimum, that is, towards the bottom of a well (see Fig. 4.1). If the energy landscape is characterised by a multiplicity of minima close together, the result may depend, to a significant extent, upon the initial conditions. Let us now increase the temperature of the system: this, in mechanical terms, corresponds to providing the material point with a disordered kinetic energy, or to "shake" the energy landscape, introducing a certain noise level. In this way, for fairly high temperatures, the probability of transition between wells is no longer negligible. At thermal equilibrium, the probability of occupation of the various wells only depends on their depth.

As we know from Chap. 4, under suitable conditions (symmetrical connections and asynchronous updating) it is possible to define an energy function for dynamical networks. In particular, in the case of a hierarchical network with a hidden layer, adopting the conventions introduced in Sect. 5.3, the energy may be written as

$$E = \sum_{i,k} W_{ki} x_i z_k + \sum_{k,j} W_{jk} z_k y_j \quad .$$

In order to simplify this expression, the term taking into account the threshold levels has not been included. However, as mentioned above, it is always possible to take into account the thresholds by introducing suitable additional units.

We can now think of introducing the analogue of a thermal agitation into a dynamical network simply by modifying the rule for updating the units. In particular, the function 1[.] in Eqs. (5.3) and (5.4) is substituted by a probability distribution.

Given that ΔE_k is the variation in energy when passing from state 1 to state 0 of the k-th unit of the hidden layer, we shall put

$$p(z_k = 1) = \frac{1}{1 + e^{-\Delta E_k/T}} \qquad (5.14)$$

and for the output layer

$$p(y_j = 1) = \frac{1}{1 + e^{-\Delta E_j/T}} \quad . \qquad (5.15)$$

Eqs. (5.14) and (5.15) reduce to (5.3) and (5.4) when the "pseudo-temperature" T goes to zero. The choice of these expressions assures that the relative probability of two global states α and β is determined only by their energy difference (see Appendix 4.1) and follows a Boltzmann distribution

$$p_\alpha/p_\beta = e^{-(E_\alpha - E_\beta)/T} \qquad (5.16)$$

The introduction of a pseudo-temperature in Eqs. (5.14), (5.15) and (5.16) necessarily implies a non-deterministic approach to the network. Consequently, the learning algorithm which will now be discussed will also be a probabilistic one.

Let us suppose that the input and output units are externally clamped; they will be generally referred to as "visible units". The vector of the visible units defined in this manner, which will be represented as V_α, will operate as a boundary condition for the rest of the network, which (if the input and the output units are clamped to the vector V_α for a sufficient length of time) will reach thermal equilibrium.

The structure of the set of vectors V_α is determined by specifying the probability distribution $P^+(V_\alpha)$, i.e. the probability distribution when the visible units are clamped. It does not depend upon the weights of the connections, but only on the characteristics of the problem.

The aim of a learning algorithm, in this case, is to modify the weights in order to bring the "free running" probability distribution $P^-(V_\alpha)$ close to $P^+(V_\alpha)$. To reach this aim, a function capable of measuring the distance between the two distributions must preliminary be defined. Such a function, in the determination of the probabilistic learning algorithm, plays an analogous role to the cost function in back propagation learning for deterministic systems.

In particular, the so called "Boltzmann machine" (Hinton and Sejnowski, 1986) is a probabilistic hierarchical network which refers to the cost function

$$G = \sum_\alpha P^+(V_\alpha) \ln \frac{P^+(V_\alpha)}{P^-(V_\alpha)} \quad . \qquad (5.17)$$

The function G is always positive and vanishes only if the two probability distributions $P^-(V_\alpha)$ and $P^+(V_\alpha)$ coincide. It is also interesting to note that, when one attempts to approximate a probability distribution, it is more important to obtain correct probabilities for frequent events rather than for rare ones. It is for this reason that the logarithm of the ratio between the probabilities is weighted, in Eq. (5.17), with the desired probabilities.

Learning thus consists of a change in the weights of the connections which reduces the value of the difference between the desired and effective probability distributions, a difference measured by Eq. (5.17). At first sight, this may seem an extremely difficult task, because a change in any single weight changes, in turn, the probabilities of the states, which also depend upon all the other weights. Fortunately, however, it is possible to find a local expression for the derivative

$$\frac{\partial G}{\partial W_{ij}}$$

that is, an expression which only depends upon the states i and j. Based upon statistical considerations, it can in fact be shown (see Appendix 5.1) that

$$\frac{\partial G}{\partial W_{ij}} = -\frac{1}{T}[p_{ij}^+ - p_{ij}^-] \qquad (5.18)$$

where p_{ij}^+ is the probability, averaged over all the vectors V_α and measured at equilibrium, that units i and j are on when the visible units are clamped, while p_{ij}^- is the corresponding probability when the network is free-running.

The choice of the pseudo-temperature T is a critical problem in probabilistic learning. At low "temperatures", in fact, a long period of time is required to reach a situation of statistical equilibrium; at high "temperatures", however, the difference between stable and metastable fixed points tends to lose its importance.

In order to tackle this problem, it is again possible to refer to a physical analogy. It is known that, in order to obtain a solid in an ordered state, corresponding to an energy minimum, an "annealing" strategy is often used, consisting of taking up the temperature to the melting point and then reducing it. The analogue of annealing in probabilistic networks is the so-called "simulated annealing", which consists of passing through a sequence of pseudo-temperatures, which are initially high and are then slowly lowered. At each one of these, a certain number of processing steps are executed, and only after the "thermal sweep" the system is allowed to relax towards statistical equilibrium.

To illustrate the operation of a "Boltzmann machine" i.e. of a hierarchical network with the previously introduced probabilistic algorithm and with simulated annealing, we will discuss two simple examples.

First of all, consider the XOR problem (Sejnowski et al., 1986), already discussed in the previous section. In agreement with Eq. (5.18), probabilistic learning was carried out in two stages, both with an annealing sequence. Each stage consists of a certain number of learning episodes.

In order to understand the method used, it is worth recalling that, as we are dealing with a homogeneous Markov (and therefore ergodic) stochastic process (Serra et al., 1986), the ensemble-average probabilities can be approximated with time-average probabilities at thermal equilibrium.

In the "+" stage, the visible units were clamped at the different possible configurations, and the hidden unit was brought to thermal equilibrium through a suitable annealing schedule.

The statistics of p_{ij}^- under free-running conditions was determined in a similar way, following the same annealing schedule.

When the learning process was based upon 2000 presentations, a success of more than 95% in the execution of the cognitive task was obtained. Table 5.1 summarises the results obtained by means of the previously defined learning algorithm. In the first column the hidden unit performs an OR, while in the second one it performs a negated OR. The other columns represent different logical functions.

Table 5.1. Summary of eight different ways in which a single hidden unit can be used in an XOR network. The desired input-output matching is shown on the left side of the table and each column on the right side shows the state of the hidden unit for a particular solution to the problem (after Sejnowski et al., 1986).

Inputs	Output	Hidden unit							
0 0	0	0	1	0	1	0	1	0	1
0 1	1	1	0	1	0	0	1	0	1
1 0	1	1	0	0	1	1	0	0	1
1 1	0	1	0	0	1	0	1	1	0
Percentage of runs:		4%	54%	16%	1%	16%	1%	7%	0.3%

As a second example of the application of a Boltzmann machine the following cognitive task (Hinton and Sejnowski, 1986) will be considered. Two 8-bit vectors V_1 and V_2 are given as input. The elements of V_1 are on or off with a probability of being on of 0.3; the vector V_2 either coincides with V_1 or is shifted by one position to the left or to the right. To avoid boundary problems, an "autowrap" is introduced so that, for example, in the case of a shift to the right, the state at the extreme right of V_1 shifts to the extreme left of V_2.

The cognitive problem to be solved in this case using a Boltzmann machine is that of recognising the relationship between the two vectors. Since there are three possibilities, it is sufficient to have an output vector consisting of three units, each of which signals the occurrence of a particular shift. The network must obviously contain a hidden layer, because this is a typical feature extraction task. As the minimum number of units in the hidden layer is a priori not known, a relatively large value (24 units) is chosen.

Also in this case, the learning procedure is a probabilistic one with simulated annealing.

Fig. 5.9 shows the weights of the connections corresponding to 24 hidden units (not mutually interconnected) after 9000 learning sweeps. Small black squares represent negative weights, while white squares represent positive ones. The dimensions of the squares represent the intensity of the weights. The lower two lines of weights correspond to the connections with V_1 and with V_2, respectively. The three weights at the centre of the upper line correspond to the connections with the output units, while the small square at the top left is the threshold of the unit considered.

The understanding of the cognitive operations carried out by the hidden units is not obvious. Moreover, it is to be expected that they reach a distributed and not a localized representation of the problem. In some cases, however, things are simpler. In particular, the first unit at the top left votes for a "shift right" if the fourth unit of V_1 and the fifth unit of V_2 are both on. Some of the 24 hidden units seem to play no role whatsoever.

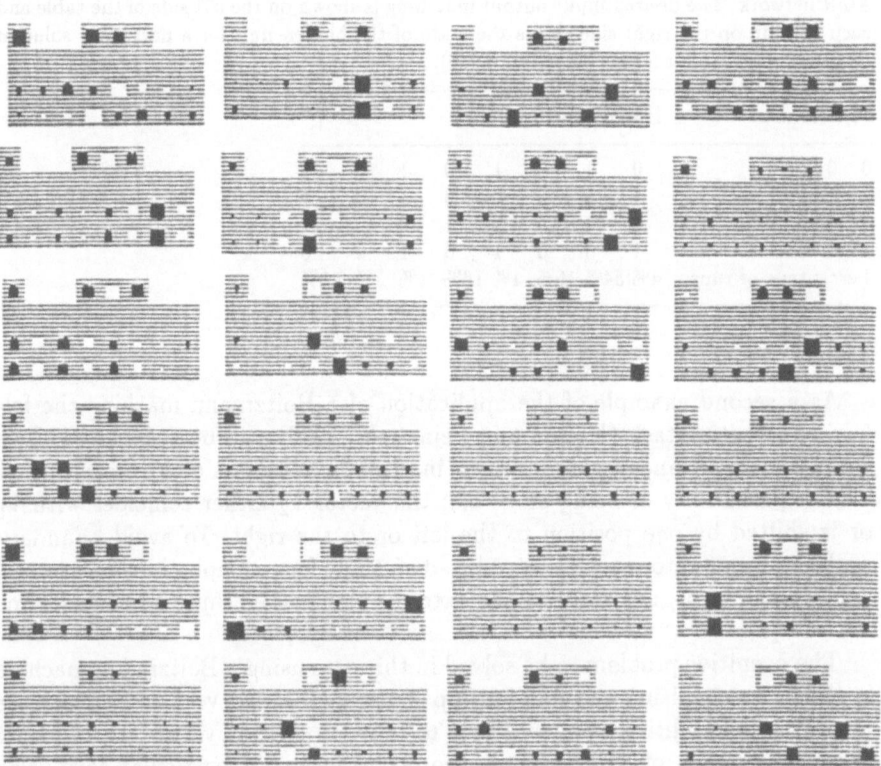

Fig. 5.9. The weights of the 24 hidden units (see text) in a shift recognizing network (after Hinton and Sejnowski, 1986)

Notwithstanding the large number of learning sweeps, the operation of this network during the recognition phase is not completely satisfactory. In particular, it always remains under 90% correctly recognised shifts.

Therefore, it is worth stressing the conceptual relevance of these networks more than than their present applicative value.

5.6 Unsupervised Learning

A brief consideration upon natural cognitive behaviours allows us to realize that the cognitive activity of feature extraction does not only occur in the presence of supervision. In particular, this happens when the categorisation is not externally defined and communicated, but is discovered (or constructed) during the analysis of the inputs.

Take, for example, word recognition by a child of suitable age. For simplicity, let us assume that four two-letter words, AA, AB, BA and BB are presented to the child a number of times, in a random sequence and with equal frequency. Normally, the child will succeed in identifying the difference between these words, thus building four classes. It can happen, however (without any external supervision), that the child "invents" a different classification, building, for example, two new classes: one made of words which begin (or end) with A, and the other made of words which begin (or end) with B (Rumelhart and Zipser, 1986).

This is an example of cognitive behaviour which can still be included in the category of feature extraction. The regularity identified, however, is not prescribed by an external teacher, but is discovered by the child through his contact with the universe of examples which he meets.

This observation leads us to explore the possibility of designing network structures which, in the absence of external supervision, are capable of inventing criteria for classifying the examples proposed.

The problem of unsupervised learning may be correctly formulated and understood only in the light of the theory of complex systems already discussed in the previous chapters (Serra et al., 1986) and, in particular, of the concept of self-organization.

In the following, we shall describe first of all the approach of competitive learning (Rumelhart and Zipser, 1986). Here, as we shall see, self-organization emerges due to the presence of interactions, nonlinearity and selective reinforcement of the connections in a multi-layered system.

The basic architecture of a competitive learning system is shown in Fig. 5.10. The connections from the input layer towards each element in the second layer are excitatory. The second layer is subdivided into clusters within which each element inhibits all the others. In other words, elements belonging to the same cluster compete with each other in responding to the input pattern. Without supervision, the validity of the response is thus entrusted to built-in selection and reinforcement mechanisms.

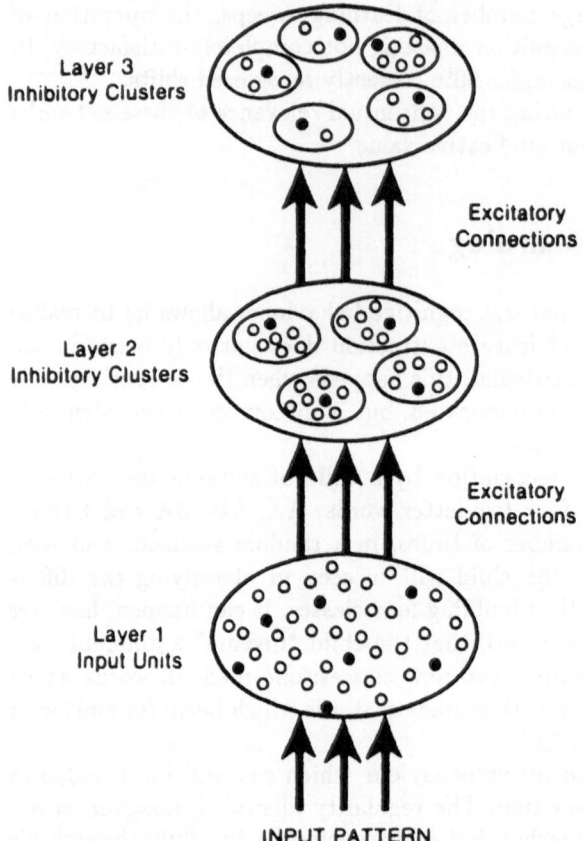

Layer 3
Inhibitory Clusters

Excitatory
Connections

Layer 2
Inhibitory Clusters

Excitatory
Connections

Layer 1
Input Units

INPUT PATTERN

Fig. 5.10. The architecture of a competitive learning system. Active units are represented by filled dots, inactive ones by open dots (after Rumelhart and Zipser, 1986)

For simplicity, only a two-layer system is described, but it would be possible to move on to hyerarchical structures with a greater number of levels.

There are various algorithms which can be adopted to implement such a competition. Here we will limit ourselves to a simple presentation of the principles, following the approach of Rumelhart and Zipser.

A unit can learn only if it can win the competition with the other units within the same cluster. Learning occurs through an increase in the active connections (associated, that is, with an active input element) and a decrease in the inactive connections.

To illustrate the operation of a competitive learning system, we will refer to the just introduced example of classification of two-letter "words". In particular, we shall quote the results of computer experiments carried out by Rumelhart and Zipser.

In one experiment, the "words" AA, AB, BA and BB were presented to a two-layer network, with one level of competing units organized in a cluster of two units. In this case, the system showed the ability to detect the relative position of the letters. In particular, one of the units spontaneously learnt to respond to the sequences AA and AB (acting as a detector of A as the leading letter) whereas the other responded to the sequence BA and BB (acting as a detector of B as the first letter). In other cases, as one might guess, the second-layer units spontaneously organized themselves to act as detectors of A or B, respectively, as the second letter.

Figure 5.11 shows the input structure for two of the four words proposed, whereas Fig. 5.12, with reference to the situation previously described, shows the active connections for each of the competing units. It can be seen that the highest weights (represented by the largest dots) correspond to the inputs common to both letters, while the lowest weights correspond to the input units which are specific to one of the two letters. The latter take the same value because the frequency with which the words are presented is roughly equal.

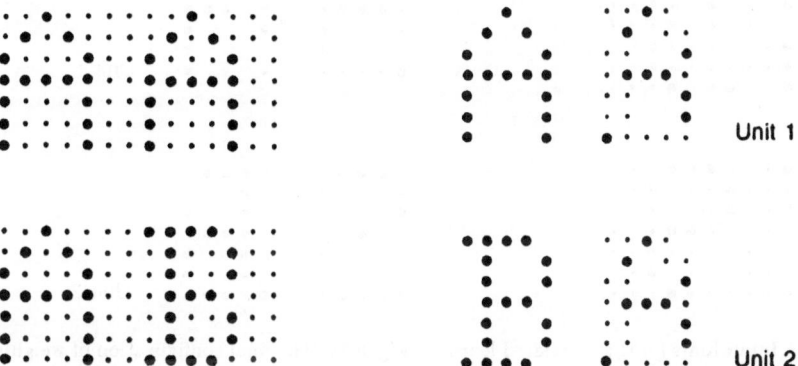

Fig. 5.11. Two input patterns (see text) to a competitive learning system (after Rumelhart and Zipser, 1986)

Fig. 5.12. The final configuration of weights for a competitive system trained on input patterns of the type shown in Fig. 5.11 (after Rumelhart and Zipser, 1986)

In a second experiment, the same stimuli were used, but the network structure was modified, increasing the number of elements in the second-layer cluster from 2 to 4. In this case, the second-layer units spontaneously learnt to behave as "word detectors", in the sense that each of them became sensitive to a specific sequence of letters.

In a further trial, the number of letters was increased while keeping the same number of letters in each word, according to the scheme AA, AB, AC, AD, BA, BB, BC and BD. In this case, if there are two elements in the second-layer cluster, they specialize in detecting the first letter of the word; if there are 4 elements, they specialize in detecting the second letter (i.e., in detecting the element which allows a grouping into four classes).

Finally, let us assume that the four words AA, BA, SB and EB are sent to the system. Obviously, there is a correlation between the first and second letter, in the sense that only A or B as the first letter can correspond to A as the second letter whereas only S or E as the first letter can correspond to B as the second letter. The letters are constructed according to the font shown in Fig. 5.13.

If the second-layer cluster contains two units, they specialize in recognising the second letter, and the weight structure which develops is that shown in Fig. 5.14. As far as the first letter is concerned, the connections corresponding to an overlap between A and B for the first detector (and between S and E for the second one) are reinforced. If, after this stage of learning, the system is only shown the letters in the first position, it will continue to associate A with B and S with E, even though the overlap of A with E (see Fig. 5.13) is greater than that of S with E.

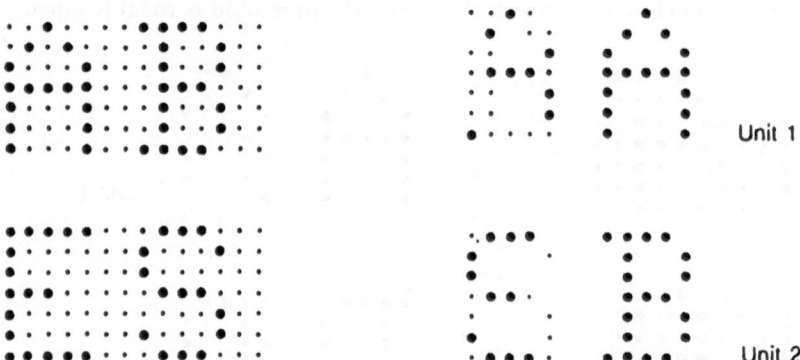

Fig. 5.13. Input fonts for the correlated learning experiment described in the text (after Rumelhart and Zipser, 1986)

Fig. 5.14. The final configuration of weights for a competitive system in the correlated learning experiment described in the text (after Rumelhart and Zipser, 1986)

In this way, a kind of conditioned learning has been obtained, showing experimentally that it is possible to influence the learning process by a competitive system (even of an unsupervised type) by controlling the statistical structure of the input.

Although these systems show limited capabilities, the most interesting point here is that of having demonstrated the possibility of the development of categorisation, and thus the emergence of a kind of cognitive behaviour in unsupervised artificial systems.

In this perspective, the very concept of error loses its meaning. When an input pattern is coded, in the sense that it leads to a specific attractor of the system, it does not, in fact, have any meaning, from the system's point of view, to ask whether the coding is right or wrong.

The problem of building cognitive systems without supervision and, in particular, of error correction without external feedback has been thoroughly studied by Grossberg (Grossberg, 1980).

The key for introducing the concept of error in an unsupervised system lies, according to Grossberg, in the internal structure of the system itself and in the communication between the various layers of the system.

As in the case of competitive learning, let us consider two interconnected layers. The first is in direct contact with the input, which determines its spatial activity pattern. In its turn, this spatial activity pattern in the first layer gives rise to an activity pattern in the second layer.

We now assume that the pattern $x_1^{(1)}$ which forms in the layer F_1 (for simplicity, this pattern is represented in Fig. 5.15a as being one-dimensional) gives rise to pattern $x_1^{(2)}$ in layer F_2.

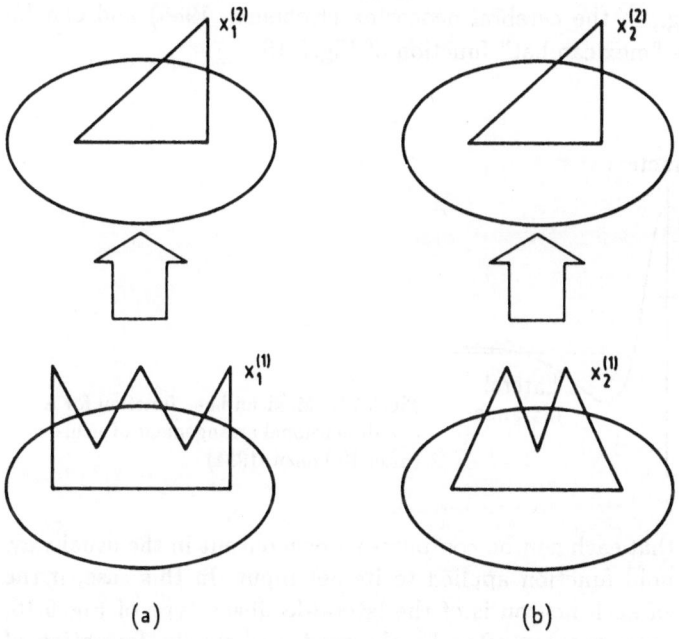

Fig. 5.15. a) pattern $x_1^{(1)}$ on the first layer elicits the correct pattern $x_1^{(2)}$ on the second layer; b) pattern $x_2^{(1)}$ elicits the incorrect pattern $x_2^{(2)}$, which is functionally equivalent to $x_1^{(2)}$ (after Grossberg, 1980)

If a pattern $x_2^{(1)}$ now gives rise in F_2 to the same pattern $x_1^{(2)}$ (Fig. 5.15b), the problem arises as to how this can be recognised as an error.

It is necessary that the network dynamics keeps track of the fact that, previously, it was $x_1^{(1)}$ and not $x_2^{(1)}$ which gave rise to $x_1^{(2)}$. More precisely, some kind of "learned feedback" from F_2 to F_1 is required: a feedback which sends to F_1 the pattern which F_2 should expect to correspond to $x_1^{(2)}$. A suitable and sufficiently rapid processing of the overlap between the two signals $x_1^{(1)}$ and $x_2^{(1)}$ must not only allow recognition of a possible error, but must also prevent the memorisation of the connection between $x_2^{(1)}$ and $x_1^{(2)}$ in case of error (and, vice versa, favour recognition through a kind of "resonance" between congruent patterns). A system of this kind, based upon learned feedback and resonance, has recently been developed (Carpenter and Grossberg, 1987).

Finally, let us discuss briefly another approach to self-organizing cognitive systems through competitive learning, which was proposed by Kohonen (1984, 1986, 1988). In order to understand the reasons behind this model, let us recall that neurophysiological evidence indicates that, in many cases, lateral feedback between neurons exists. This means that synaptic coupling from a given neuron to its neighbours is excitatory for all those neurons whose distance is smaller than a certain critical radius, while it is inhibitory for those neurons which lie at a greater distance. At even greater distances, the coupling is again weakly excitatory or negligible. This kind of interaction is fairly common, e.g., in the cerebral neocortex (Kohonen, 1986) and can be represented by the "mexican hat" function of Fig. 5.16.

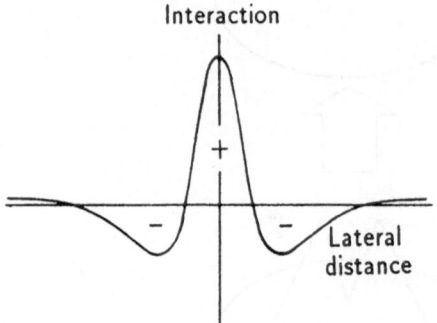

Fig. 5.16. "Mexican hat" function for a one-dimensional arrangement of neurons (after Kohonen, 1984)

Let us suppose that each neuron computes its own output in the usual way, i.e. through a sigmoid function applied to its net input. In this case, if the synaptic coupling of each neuron is of the lateral-feedback type of Fig. 5.16, an interesting phenomenon can often be observed, namely the formation of activation clusters, or "bubbles". If the initial activation values are spread in a wide portion of the network, it tends to concentrate within a bubble of limited radius, where activation reaches high values, while it vanishes outside the cluster (Fig. 5.17).

This observation is the starting point of Kohonen's model. In this model, in order to simplify the computation, instead of adopting an interconnection matrix of the "mexican hat" type, a simpler equation is preferred, which explicitly takes into account the clustering phenomenon. We will now qualitatively describe the main features of this "short learning algorithm". This qualitative description will be followed by a more precise, quantitative one.

Let us suppose that all the neurons belonging to the network receive the same external input, which is in general a multi-dimensional one. Each component of this signal is multiplied by a coupling coefficient which is analogous to a synaptic coefficient, but which connects neurons with input. These products

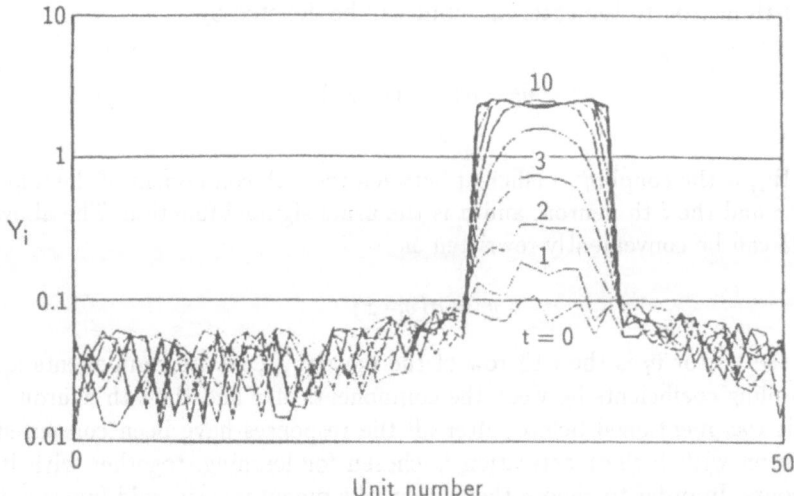

Fig. 5.17. Time sequence of the outputs of a one-dimensional neural network with a lateral coupling matrix defined by a "mexican hat" function (after Kohonen, 1984)

are then summed, in order to compute the usual "net input", or "local field" for every neuron. The coupling coefficients of the input to the various neurons differ from each other, so that every neuron reaches a different activation value. A competition then sets in, like in other competitive learning models, and the neuron with highest activation wins. The connections between the winner and the external input are then modified in such a way as to further enhance its response to that stimulus. In this way, neurons become progressively specialized in responding to particular inputs. Since the net input to a given neuron is the weighted sum of its incoming signals, these specialized neurons can classify the various input patterns into classes.

So far, we have considered only the coupling between neurons and external input. The effect of the interconnections between neurons is modelled as a clustering phenomenon: not only the winner modifies its weights in order to become more sensible to the incoming signal, but also all its neighbouring neurons (i.e., those which belong to a circular neighbourhood of predefined radius) undergo a similar adaption. While initially a homogeneous network was intruduced, as learning proceeds one can observe a progressive specialization of parts of the network, as a result of the self-organization properties of the algorithm.

Let us now describe the algorithm precisely (Kohonen, 1986). The input vector, which is sent in parallel to all the neurons, will be denoted by

$$x = (x_1, x_2, \ldots, x_n) \quad .$$

The neurons will be assumed to form a two-dimensional array; for simplicity, however, each neuron will be denoted by a single subscript. The response of the i-th neuron to the external input will be denoted by

$$y_i = \sigma\left(\sum_{j=1}^{n} W_{ij}x_j\right)$$

where W_{ij} is the coupling coefficient between the j-th component of the input vector x and the i-th neuron, and σ is the usual sigmoid function. The above formula can be conveniently rewritten as

$$y_i = \sigma(v_i \cdot x)$$

where the vector v_i is the i-th row of the matrix W, i.e. its components are the coupling coefficients between the components of x and the i-th neuron.

As it was mentioned before, after all the responses have been computed, the neuron with highest activation is chosen for learning, together with its neighbours. In order to choose the winner, the monotonic sigmoid function is obviously inessential, and it suffices to compute the maximum of the scalar product:

$$\max_{i}(v_i \cdot x) \quad .$$

Let us now suppose that the k-th neuron is the winner of the competition, and let E_k be a circular neighbourhood of a given radius, centered around it.

Item	A	B	C	D	E	F	G	H	I	J	K	L	M	N	O	P	Q	R	S	T	U	V	W	X	Y	Z	1	2	3	4	5	6
Attribute																																
a_1	1	2	3	4	5	3	3	3	3	3	3	3	3	3	3	3	3	3	3	3	3	3	3	3	3	3	3	3	3	3	3	3
a_2	0	0	0	0	0	1	2	3	4	5	3	3	3	3	3	3	3	3	3	3	3	3	3	3	3	3	3	3	3	3	3	3
a_3	0	0	0	0	0	0	0	0	0	1	2	3	4	5	6	7	8	3	3	3	3	6	6	6	6	6	6	6	6	6	6	6
a_4	0	0	0	0	0	0	0	0	0	0	0	0	0	0	0	0	0	1	2	3	4	1	2	3	4	2	2	2	2	2	2	2
(a) a_5	0	0	0	0	0	0	0	0	0	0	0	0	0	0	0	0	0	0	0	0	0	0	0	0	0	0	1	2	3	4	5	6

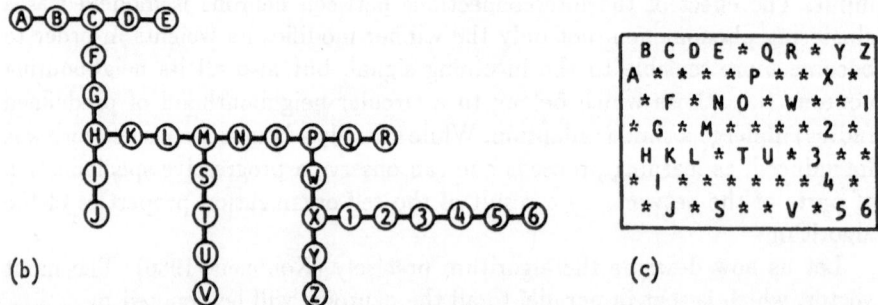

Fig. 5.18a-c. A self-organizing neural network, applied to a symbol recognition task. (a) Letters and numbers are coded with a five-dimensional input vector; (b) a minimal spanning tree of the input data, representing the similarities among the data; (c) a representation of the network after learning: to each input item the neuron is associated with the highest response. Note that similar vectors are associated to neighbouring neurons

Let us call Δv_j the modification of the weight vector v_j after presentation of the input pattern x. The learning algorithm modifies the weights according to the following rule:

$$\Delta v_i = \begin{cases} 0 & \text{if } i \notin E_k \\ \alpha(x - v_i) & \text{if } i \in E_k \end{cases}$$

where $\alpha < 1$ is a positive adaptive constant.

Kohonen has shown several applications of his model to different problems (for an example, see Fig. 5.18).

The learning algorithm presented so far defines its own categorization of different input patterns into classes through its self-organization properties. A modification of the original model, which allows the external teaching of the correspondence between patterns and classes, has been recently introduced (Kohonen, 1988). This so-called "learning vector quantization" algorithm, is finding some very interesting and promising applications. e.g. in speech recognition.

Appendix A5.1 The Learning Algorithm for the Boltzmann Machine

This Appendix derives the learning algorithm (Eq. (5.18)), following the procedure proposed by Hinton and Sejnowski (Hinton and Sejnowski, 1986).

First of all, we note that when the network is free running and has reached thermal equilibrium, the probability distribution of the visible units can be expressed as

$$P^-(V_\alpha) = \sum_\beta P^-(V_\alpha, H_\beta) = \frac{\sum_\beta e^{-E_{\alpha\beta}/T}}{\sum_{\lambda\mu} e^{-E_{\lambda\mu}/T}} \qquad (A5.1)$$

where V_α is the state vector of the visible units, H_β is the state vector of the hidden units and $E_{\alpha\beta}$ is the corresponding overall energy.

Adopting, for simplicity, a numeration of the units which is layer-independent, the expression for the energy may be rewritten as

$$E_{\alpha\beta} = -\sum_{1<j} W_{ij} s_j^{\alpha\beta} s_i^{\alpha\beta} \qquad (A5.2)$$

Differentiating gives

$$\frac{\partial E_{\alpha\beta}}{\partial W_{ij}} = \frac{1}{T} s_j^{\alpha\beta} s_i^{\alpha\beta} e^{-E_{\alpha\beta}/T} \qquad (A5.3)$$

Differentiating $P^-(V_\alpha)$ with respect to W_{ij} and substituting, we have,

$$\frac{\partial P^-(V_\alpha)}{\partial W_{ij}} = \frac{1}{T}\left[\sum_\beta P^-(V_\alpha, H_\beta)s_j^{\alpha\beta}s_i^{\alpha\beta}\right.$$
$$\left. - P^-(V_\lambda, H_\mu)\sum_{\lambda\mu} P^-(V_\lambda, H_\mu)s_j^{\lambda\mu}s_i^{\lambda\mu}\right] \quad .$$

(A5.4)

This derivative, in its turn, allows an evaluation of the gradient of the function

$$G = \sum_\alpha P^+(V_\alpha)\ln\frac{P^-(V_\alpha)}{P^+(V_\alpha)}$$

(A5.5)

recalling that $P^+(V_\alpha)$ does not depend upon W_{ij}. In fact, one then obtains

$$\frac{\partial G}{\partial W_{ij}} = -\sum_\alpha \frac{P^+(V_\alpha)}{P^-(V_\alpha)}\frac{\partial P^-(V_\alpha)}{\partial W_{ij}}$$

$$= -\frac{1}{T}\left[\sum_\alpha \frac{P^+(V_\alpha)}{P^-(V_\alpha)}\sum_\beta P^-(V_\alpha, H_\beta)s_j^{\alpha\beta}s_i^{\alpha\beta}\right.$$
$$\left. - \sum_\alpha \frac{P^+(V_\alpha)}{P^-(V_\alpha)}P^-(V_\alpha)\sum_{\lambda\mu} P^-(V_\alpha, H_\beta)s_j^{\lambda\mu}s_i^{\lambda\mu}\right] \quad .$$

(A5.6)

We now recall that one may write

$$P^+(V_\alpha, H_\beta) = P^+(H_\beta, V_\alpha) = P^+(H_\beta \mid V_\alpha)P^+(V_\alpha)$$
$$P^-(V_\alpha, H_\beta) = P^-(H_\beta, V_\alpha) = P^-(H_\beta, V_\alpha)P^-(V_\alpha) \quad .$$

(A5.7)

Moreover, it must be

$$P^-(H_\beta, V_\alpha) = P^+(H_\beta, V_\alpha)$$

(A5.8)

because the probability of a certain vector of the hidden states, given the vector of the visible units, must not depend upon the way in which these units have reached their particular states.

On substituting this equation in the preceding expression and recalling that

$$\sum_\alpha P^+(V_\alpha) = 1$$

(A5.9)

one finally obtains

$$\frac{\partial G}{\partial W_{ij}} = -\frac{1}{T}(p_{ij}^+ - p_{ij}^-)$$

(A5.10)

where

$$p_{ij}^+ = \sum_{\alpha\beta} P^+(V_\alpha, H_\beta)s_j^{\alpha\beta}s_i^{\alpha\beta}$$

$$p_{ij}^- = \sum_{\lambda\mu} P^-(V_\lambda, H_\mu)s_j^{\lambda\mu}s_i^{\lambda\mu} \quad .$$

(A5.11)

6. Dynamical Rule Based Systems

6.1 Introduction

In the second chapter of this volume we have put forward the idea that complex dynamical systems are capable of carrying out cognitive tasks, and we have shown the importance of studying artificial cognitive systems for a comprehension of the general characteristics of cognitive processes. The following chapters illustrate these concepts discussing the various approaches and models.

Most of the dynamical cognitive systems examined so far belong to the category of neural networks, characterized by the fact that the evolution law of any single element ("formal neuron") takes its inspiration from knowledge of the dynamics of biological neurons.

The fact that the brain is an extraordinary cognitive system is undoubtedly an element in favour of choosing models of this type. Moreover, one may hope that some of the knowledge acquired about artificial neural systems could be important for a better understanding of the functioning of the human brain. We must, however, remember that the human brain is much more complex than any artificial system so far proposed, and that knowledge of its operation is still very limited.

On the other hand, we have also discussed non-neural models capable of interesting behaviour such as, for example, reaction-diffusion models and random boolean networks (Chap. 3).

A fundamental difficulty of all the models examined so far, however, is linked to the introduction of domain specific knowledge. In fact, it is unrealistic to think that a completely generic network, like those examined so far, is capable of learning complex tasks only from a series of examples, without any form of a priori knowledge. One may consider human learning to understand how many things are explicitly taught, instead of being discovered by trial and error.

In other words, it is highly improbable that a system starting from scratch can go very far. This important point, only mentioned here, will be taken up again in Chap. 7, within the examination of the limits of the dynamical approach to artificial intelligence.

On the contrary, a priori domain knowledge can be directly introduced in the so-called "classifier systems" (Holland, 1986; Holland et al., 1986), which are nonlinear dynamical systems also capable of learning by examples. One can thus propose to develop a complex cognitive system in two stages, the first one making use of a limited amount of "knowledge engineering", while the capabilities of the system are augmented through automatic learning in the second stage.

Besides providing a simple solution to the problem of introducing a priori knowledge, classifier systems constitute an interesting link between cognitive dynamical systems and classical AI. They are, in fact, based upon the use of production rules, called "classifiers": domain knowledge can therefore be introduced using usual knowledge engineering techniques. Each classifier is associated with a numerical variable, its "strength", and the evolution law of the strengths defines a nonlinear dynamical system.

It is interesting to note that also classifier systems, like neural networks, are inspired by biology. This time, however, the source of inspiration is not human brain, but biological evolution. Indeed, natural evolution may be considered a form of learning, since it is based upon modifications reflecting the characteristics of the environment. One could say that any living organism constitutes a "memorization" (or better, an "interpretation") of the characteristics of the environment in which its ancestors lived.

The mechanisms of mutation are, to a large degree, random, while the environment takes on the task of selecting the most successful mutants, i.e., those with the greatest fitness.

As we shall see, the original formulation of classifier systems is directly inspired by a knowledge of the mechanisms of genetic recombination, and the classifiers are treated in an analogous manner to chromosomes. The existence of a close analogy between classifier systems and the immune system has also been stressed. The dynamical equations of the former have, in fact, a form similar to that of the evolution equations for antibody populations, obtained from Jerne's idiotypic theory (Farmer et al., 1986).

Also, from an epistemological and history of science point of view, it is interesting to note that all dynamical cognitive systems studied so far are based upon – and have, in effect, been thought of stemming from – analogies with natural systems, nearly always biological ones (human brain, chromosomes, immune system) and occasionally physical ones (spin glasses, chemical reactions).

Even the humanities (economics, sociology) have provided useful metaphors for the development of artificial cognitive systems, such as in Minsky's (1986) theory of the "society of mind". We shall find an example of the role played by these metaphors also in the field of classifier systems, where the model adopted to determine the fitness of a particular individual (the "bucket brigade" algorithm) is inspired by simple economic reasoning.

In the spirit of this volume, the study of classifier systems does not aim to describe natural or social phenomena faithfully. It rather draws inspiration from the (incomplete and partial) knowledge which we have of them, in order to construct abstract models of systems capable of learning.

Every model of this type must clearly define i) the selection mechanisms, within a given population and ii) the mutation mechanisms of that population. Although in biological systems these mechanisms are simultaneously active, their effects manifest themselves on different time scales. We can identify at least three different time constants: the first one describes the velocity of selection of the fittest individual (more precisely, the relative rate of reproduction of the most fertile one) within a fixed population. The second time constant describes the rapidity with which mutations are generated. Finally, there is the time scale of environmental variations, which describes changes in the definition of the fitness of individuals.

In the case of artificial systems which can learn, one must also specify a mechanism whereby the "fitness" of different individuals (i.e., the correctness of what they have learnt) is evaluated. In classifier systems, as is often the case in artificial cognitive systems, the role of the external environment is played by a teacher who provides the system with information about the suitability of its answers.

The next section will introduce general ideas about classifier systems, whereas the algorithms will be defined in detail in Sect. 6.3. As we have already noted, classifier systems are dynamical systems and their dynamical properties will be analyzed in Sect. 6.4, reference being made to a particular "cognitive task", i.e. the forecasting of sequences of letters. Finally, the last section of this chapter is a more detailed examination of the relationships between classifier systems and neural networks.

6.2 Classifier Systems and Genetic Algorithms

Classifier systems were introduced by Holland (1986) in an attempt to apply to cognitive tasks the so-called genetic algorithms (Holland, 1975), which had been used, above all, to solve complex optimization problems (Grefenstette, 1986 and 1987; Davis, 1987; Muehlenbein, 1988).

The fundamental idea behind genetic algorithms is that of introducing into the system under study a mutation mechanism and an external selective pressure, which favours high fitness mutants, eliminating the least adapted ones.

Classifier systems are similar to production systems in artificial intelligence, since they are essentially based upon rules of the "if ... then" type. Classifier systems thus provide a link between the classical AI approach and the dynamical approach. These aspects will be examined in more detail in the discussion on the relationships between the classical and the dynamical approach, in Chap. 7.

Let us now examine classifier systems in detail. Since we are dealing with rather complicated systems, with numerous variants, we will begin by describing a particularly simple basic version, and then go on to discuss some of the most important variants.

A classifier system (Fig. 6.1) consists of:

1) A set of detectors (input devices) which provide information to the system about the state of the external environment.
2) A set of effectors (output devices), which transmit to the external environment the conclusions reached by the classifier system.
3) A set $C = \{C_i\}$ of rules, called classifiers; each classifier C_i consists of a condition c_i and an action $a_i : C_i = [c_i/a_i]$.
4) A list of active messages $ML = \{m_j\}$.

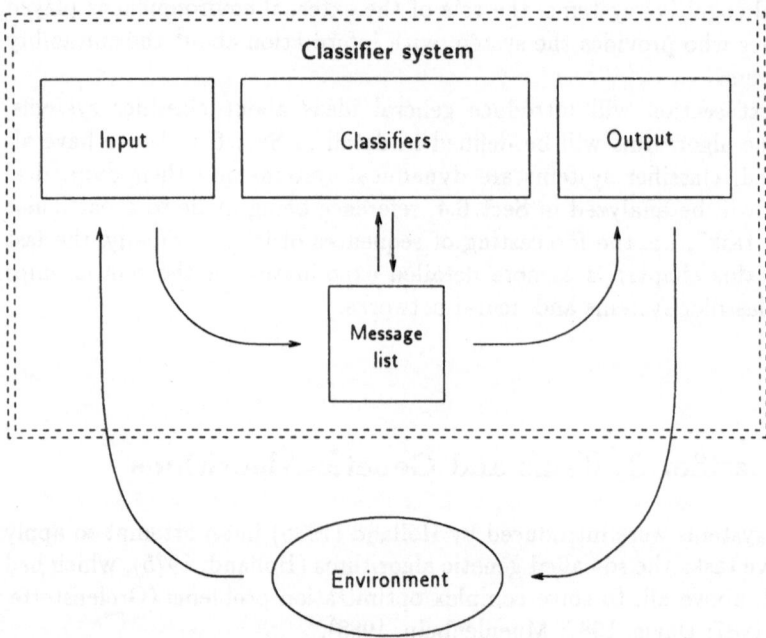

Fig. 6.1. Information flows in classifier systems

Learning is supervised as in multilayered networks. The external environment provides information to the system through its detectors. This information is processed by the system until the latter produces an answer, which is then communicated to the external environment by the effectors. At this point, the teacher provides the system with feedback about the correctness of the answer. In a multilayered network, the detectors would correspond to the units in the input layer, and the effectors to those in the output layer.

A classifier system works with messages: the detectors translate the information about the external world in the form of messages, which are placed in the message list. The effectors check, at each time step, whether output messages are present and, if this is the case, they proceed to send them to the external environment.

Classifiers, as mentioned above, are composed of two parts, a condition and an action. The former defines under which circumstances the classifier should be activated, while the latter defines what it should do. In practice, in each cycle, each classifier verifies whether the message list contains one or more messages which "match" its condition part (in a sense defined below). If this is the case, the classifier is activated and tries to post a new message, specified by its action part.

Not every classifier which succeeds in matching with a message, however, is successful in posting its own message. Each classifier C_i is, in fact, associated with a numerical variable s_i which measures its "strength", and which is used to determine which classifiers will actually post their own message. Thus, there is competition between classifiers which have found a match and the "strongest" ones prevail.

This competition can take place either by imposing a constraint upon the maximum number of messages which can be posted at any given time, or by fixing a minimum acceptable strength, so that classifiers whose strength is lower than the threshold are excluded from posting messages (the threshold itself might depend upon other variables such as, e.g., the number of matching classifiers).

Actually, as we shall see below (Sect. 6.3), the competition is not based directly upon strengths, but upon particular functions of the classifiers' strengths called "bids". Moreover, the competition is ruled by a stochastic mechanism, so that the probability of winning is determined by the bid: this randomness plays a role similar to that of finite temperature in simulated annealing, that is it sometimes allows the testing of unfavourable alternatives, discouraging them without imposing absolute inhibition.

In the implementations of classifier systems developed so far (Riolo, 1988), the messages are coded, in a somewhat primitive manner, as fixed-length (n) strings of symbols taken from the alphabet $\{0, 1\}$. Classifiers are composed of two parts, each having the same length as the messages: both the condition and the action are composed of strings of n symbols taken from the alphabet $\{0, 1, \#\}$. A condition c_i is said to match a given message m_k if every symbol 0 or 1 in c_i corresponds to the same symbol in m_k. The symbol "$\#$" means "don't care": in other words, the message-condition match is not influenced by the portion of the message in positions corresponding to this symbol in the condition.

In the case where the match occurs, the message generated, m_j, will consist of the following: in those positions where the action has the symbols "0" or "1", there will be that same symbol, while in those positions where the action has the symbol "$\#$" (which in this case has the meaning "pass through", and not "don't care"), the message m_j will contain the symbol found in the corresponding position in the condition; in the case where this is a "don't care" symbol, the symbol in the corresponding position in the "activator" message m_k will be chosen.

In order to exemplify what has been said above, suppose that the following message is found in the list:

$$10110$$

and that there are two classifiers:

$$10\#\#\# / 001\#\#$$
$$100\#\# / 1\#\#\#\#$$

The first one matches, while the second one does not. The message generated by the first classifier will then be:

$$00110$$

It is important to notice that, although so far we have only considered classifiers with a single condition and a single action, it is often convenient to use classifiers with multiple conditions. A classifier of this type could post only if all its conditions are satisfied by messages contained in the message list.

In the case of multiple conditions, the "matching" mechanism remains the same, whereas the method of activating the "pass through" mode has to be redefined. The simplest choice consists in placing the symbols of the message which activated the first condition in place of the "#" symbols in the action. For example, in the case where the two-condition classifier

$$11\#\#\#, 10\#\#\# / 111\#\#$$

is present and the messages which match the conditions are, respectively

$$11010$$
$$10001$$

the message posted will be

$$11110$$

Obviously, other choices are possible, like for example logical combinations of the variables of the messages which satisfy the condition parts (Riolo, 1986). If, for instance, we choose the AND of the messages, in the preceding case we obtain the message 11100, whereas if we choose the OR we have 11111.

Messages posted by classifiers can be output messages, recognized by the effectors and transferred to the external environment, or they may be internal messages to be used only by other classifiers. The idea is to generate chains of classifiers capable of carrying out useful computations and to privilege them with respect to the other classifiers.

There are two learning mechanisms which are encharged with reaching this objective. One algorithm, which operates with shorter time constants, has the aim of defining an adequate distribution of the strengths within a given population of classifiers. In this way the most useful classifiers, or chains of classifiers, within a given set of rules can be brought to prominence. The strength of a classifier is thus a measure of its utility (demonstrated, in the past, by its carrying out assigned tasks): a classifier's strength can therefore be considered as a measure of its "fitness". There are several possible alternative algorithms: the one commonly used in classifier systems is called "bucket-brigade" algorithm, and will be described in Sect. 6.3.

The second learning mechanism operates over a rather longer time scale, and modifies the classifier population. In fact, the bucket-brigade algorithm can highlight the classifiers which are most useful to the system but, if the initial population does not already contain the rules necessary for carrying out a given task, it cannot "invent" them. The creation of new useful classifiers, however, is precisely the job of the so-called "genetic algorithms". More precisely, random combinations of pre-existing classifiers are created through the use of genetic operators, privileging those with the greatest strength. The new classifiers then substitute the weaker ones, and enter into competition with the others through the bucket-brigade mechanism. This method possesses two advantages: if a mutant shows a high level of utility, its strength gradually increases and it becomes a "strong" classifier, until it may replace a pre-existing classifier which carries out a function which is similar but in a less appropriate manner. In addition, the overall properties of the system should not be destroyed by genetics: the new classifiers, in fact, substitute the weaker representatives of the previous population, and it is reasonable to suppose that the latter contributed very little to the overall performance of the system. Naturally, for this mechanism to be effective, the genetics has to be slow enough to allow a "thermalization" of the set of rules, so that a classifier's strength is a meaningful measure of its utility for performing the given task, and does not merely reflect some random fluctuation.

Various genetic operators can be used to obtain new classifiers from the preceding ones. The operators described here draw their inspiration from biological genetics: the analogy between the two systems is based upon the interpretation of both classifiers and chromosomes as strings of symbols.

The simplest genetic operator is "point mutation": the value of a symbol in a randomly chosen position is modified in a random manner. The new classifier is then only slightly different from its parent, which in this case is unique. Note that some mutations can be interpreted intuitively: for example, substituting a "1" or a "0" with a "#" corresponds to a limited generalization of the preceding rule, while the inverse mutation is a specialization.

Another genetic operator which can be used is the "inversion" operator. Also in this case, the parent is unique. Two positions are selected in the string: the substring included between these two positions is then inverted. Let us indicate two substrings as "A" and "B"; once the two positions have been selected, if the parent classifier is, for example, $A.11000.B$, then the "child" obtained through inversion is $A.00011.B$.

The most widely used genetic operator, however, is the "crossover" operator. Two "children" are obtained from two different parents by taking a cut-point and exchanging their segments. To make this clearer, suppose that there are two classifiers C_1 and C_2, and a position is chosen at random along the string which represents them. The semi-strings found can be represented by letters separated by a dot:

$$C_1 = A.B$$
$$C_2 = C.D$$

The crossover operator then generates two new classifiers, which can be represented as:

$$C_3 = A.D$$
$$C_4 = C.B$$

There is also a variant of the crossover operator, the double crossover, which exchanges an internal substring between two classifiers. For example, taking two classifiers:

$$C_1 = A.B.C$$
$$C_2 = D.E.F$$

we obtain:

$$C_3 = A.E.C$$
$$C_4 = D.B.F$$

The crossover operator has the effect of generating new classifiers, using building blocks already present in the parents, but combining them in a different manner. In this way it is possible to explore new alternatives, using portions of the structures which proved to be effective in the past (recall that the parents are preferentially chosen amongst the high strength classifiers). The crossover operator plays a decisive role in the property which these systems possess and which is known as "implicit parallelism". In order to clarify this concept, the notion of "schema" must be defined.

Considering a classifier as a string of symbols, a "schema" is associated to a substring, composed of symbols which are not necessarily contiguous. Each schema identifies a subset of the set of classifiers (a hyperplane), defined by the fact that all its members contain that particular substring. For example, a schema can be defined by the set of all classifiers with a "1" in the first position (1*****), another by all those with a "1" in the first position and a "0" in the last position (1****0), etc.

The symbol "*" is used to indicate those positions where the corresponding symbol does not take part in the definition of the schema: it should not be confused with the "don't care" symbol.

We can thus say that every schema provides its own contribution to the overall performance of the classifiers, and that the problem is to succeed in testing a sufficiently high number of schemata in a reasonable period of time.

Holland has shown that genetic algorithms are capable of sampling the space of possible schemata in an efficient manner ("implicit parallelism"). The number of possible schemata increases exponentially with the length of the string and an explicit evaluation of the performance of the various schemata would require an enormous amount of time. Genetic algorithms carry out an implicit exploration of the "schemata space", privileging in successive generations those schemata showing the best performance.

The most important genetic operator in this perspective is the crossover operator, while point mutations essentially play the role of guaranteeing a comprehensive exploration of the phase space. In fact, on starting from an arbitrary initial classifier population, there is no guarantee that all the 3^{2n} possible classifiers can be generated by crossover. This can easily be seen by considering, for example, only classifiers which begin with the symbol "1": no other symbol can be present in that position, using only the crossover operator. The addition of mutations, on the other hand, guarantees that the system has the possibility of exploring the entire phase space.

A mathematical description of implicit parallelism is given in Appendix 6.1.

In order to apply the genetic operators described above, one must define how the parent classifiers are found, and how those which are to be eliminated are chosen. This choice is based upon the strength of the classifiers, usually in a random way: that is, a probability distribution of being selected for parenthood is defined as an increasing function of the classifiers strengths, and parents are picked up according to this probablity. A similar method is used for finding the classifiers to be replaced, although of course, in this case, the probability is a decreasing function of the strength.

As observed by Robertson and Riolo (1988), in some cases the use of "classical" genetic operators only (i.e., those described above) does not allow satisfactory levels of performance. Several other genetic operators have therefore been introduced. It should also be stressed that these operators are of general interest and are not ad hoc operators whose use is limited to only one or very few kinds of tasks. They are known as "cover detector", "cover effector" and "triggered chaining" operators.

The first one enters into action in the case where an input message from the detectors is present, but there is no classifier which matches it. Let us take, for example, the situation at the outset, with a completely random population of classifiers. The probability of a "no-match" situation is high, carrying the consequence that the reward/punishment mechanisms of the bucket brigade are unable to operate. The "blind" genetics described above would carry out an exploration of possible alternatives, without any help from the selective pressure which privileges, as parents, the strongest classifiers. This blind exploration, however, requires very long times and, in the meantime, the average strength of the population may decay due to the effect of dissipative terms.

The cover detector mechanism allows this step to be shortened: every time there is no classifier which matches an external message, it generates a classifier which does match and has a random action. In this way the selective evaluation mechanisms of the bucket brigade can enter into operation.

The cover effector mechanism, on the other hand, enters into action when the system is unable to generate an output, or when it generates a wrong one. This activates a genetic mechanism which alters, in a random manner, the action of one or more classifiers, chosen amongst those which are capable of posting a message in that situation.

The combined effect of introducing cover mechanisms is to increase, by several orders of magnitude, the learning rate (Robertson and Riolo, 1988; Compiani et al., 1989a).

The triggered chaining operator also carries out an interesting task: it tries to create classifier chains. Two classifiers are chained if one of them posts a message which is exploited by the other. The triggered chaining operator is activated if, at time t, there is at least one classifier (say C_2) which gains a good profit (increasing its own force by a value greater than a certain threshold). In this case, it chooses one of the classifiers which have posted a message at the preceding point in time (say C_1), and generates two new classifiers (C_3 and C_4): the condition part of the former is equal to that of C_1, while the action part of the latter is equal to that of C_2. Moreover, the action of C_3 is helpful in activating C_4.

For example, let us consider classifiers with two conditions, like those of the CFS-C system described in Sect. 6.4. In this case, the triggered chaining operator works as follows: let the parent classifiers be

$$C_1 : c_{11}, c_{12}/a_1$$

$$C_2 : c_{21}, c_{22}/a_2$$

In this case, the triggered chaining generates two classifiers of the type:

$$C_3 : c_{11}, c_{12}/a'$$

$$C_4 : c_{21}, a'/a_2$$

Therefore, this operator tends to help in creating situations where, if the conditions of C_1 are satisfied at time t, the action of C_2 is posted at the next time step, that is, it tends to reproduce an association which proved useful in the past. This genetic operator can be regarded as an analogue of the Hebb hypothesis, which reinforces the coupling between two classifiers which not only fired one after another, but did it in a rewarding manner. The triggered chaining operator carries out an important function in the development of spontaneous memorization mechanisms (see below).

6.3 The Equations of Classifier Systems

In the previous section we described the main characteristics of classifier systems and genetic operators. The evolution laws, or "equations of motion", of classifier systems will now be precisely defined.

The system carries out a sequence of "major cycles", which will be numbered as $t, t+1, \ldots$ The transition between the state at time t and the successive one involves the following steps.

Step 1: The detectors check whether input messages are present and, if so, they are posted in the "current message list" ML_1. This list also includes messages posted by the classifiers in the preceding cycle. The "new message list" ML_2 is then emptied.

Step 2: All the messages are compared with the conditions of the classifiers.

Step 3: All the classifiers whose conditions are matched "join the race" to post messages. The race is based upon a bid, which is a function of the strength of the classifier (the formulae are given below). The winners, which are chosen according to a probability distribution which favours the highest bidders, "become active", i.e., they post their own message(s) in the "new message list" ML_2. Due to the "don't care" mechanism, a classifier may match and post more than one message.

Step 4: Classifiers winning the competition "pay" the classifiers which had previously posted the messages which they match. This payment involves a reduction in the strength of the classifier which has become active in favour of that which allowed it to do so.

Step 5: The old message list, ML_1, is cancelled and substituted by the new list ML_2 which contains the messages posted by the winning classifiers.

Step 6: The effectors check whether there are output messages present in ML_2. If they are present, they are suitably coded and sent to the external environment: the teacher then provides the system with feedback, in the form of a reward or a punishment to the classifier which generated the output. This reward is paid for in the system's currency, i.e., by an increase or a decrease in the strength. In the case where there are no output messages, the system proceeds to the next step.

Step 7: Eventually, genetic operators are applied which modify the classifier population. Usually, in order to allow a thermalization of the rules, the genetic operators are not applied at every major cycle: for example, their activation may be a probabilistic one, with an average frequency of $1/T$ steps. The probability that a classifier is chosen as a parent is an increasing function of its strength, while its probability of being replaced by a newborn is a decreasing function of its strength. Protections can be introduced in order to prevent the elimination of some classifiers (e.g., those whose strengths are above average).

Step 8: Return to step 1.

The competition and payment mechanisms remain to be precisely defined. The fundamental idea is to select chains of classifiers capable of leading to appropriate answers. A classifier C_t, which directly generates an output, is rewarded by the teacher, and experiences an increase or a decrease in its own strength according to the appropriateness of its action. The difficult part is deciding how to pay the "suppliers" of C_t, that is, those classifiers which generated the sequence of messages which led to the activation of C_t.

One could trace all the classifiers which contributed to the generation of the chain, and reward or punish all of them according to the evaluation of the result. However, this direct strategy comes across memory problems, which grow with the length of the chain. It is greatly preferable to have a "local" learning rule in the sense that it involves only a classifier C_t and its "supplier" C_{t-1}, which posted the message with which C_t matches. The definition of a correct back propagation mechanism for the teacher's rewards constitutes the so-called credit assignment problem.

The bucket-brigade algorithm, which has been proposed for its solution, draws inspiration from the metaphor which sees classifiers as middlemen in a trading economy. They are "economic subjects" which try to post messages and, in order to succeed, pay a bid. If their message then turns out to be useful, they are paid, in their turn, by those classifiers which use it. In this way classifiers which post unusable messages see a decline in their strengths, while those which post useful messages are repaid, either directly by the external environment, or by other classifiers.

The bid b_i of the i-th classifier C_i, in its simplest version, is equal to a fraction of its strength s_i:

$$b_i = \alpha s_i \tag{6.1}$$

with $\alpha < 1$.

Holland (1986) proposed a slightly different formula for the bid, which privileges, in the competition for posting messages, the most specific classifiers. The specificity γ of a classifier is defined as the ratio between the number of positions in the condition occupied by symbols other than "don't care" ones and the total number of positions available. The reasons behind the choice of favouring the specific classifiers are discussed below. The modified Holland formula is

$$b_i = \alpha \gamma_i s_i \tag{6.2}$$

If only one classifier has posted the message which the new classifier matches, the entire amount of the bid has to be paid to a single supplier. It is, however, possible that a classifier is activated by messages posted by several classifiers. The latters, in this case, share the bid paid by the former.

It has been pointed out by Robertson and Riolo (1988) that sharing the bid among all the "parents" of a given classifier may render it difficult to create long chains of classifiers, if we are dealing with multiple-conditions classifiers. In this case, the different classifiers which posted the messages which match the different conditions also share the bid: each of them receives therefore only a fraction of the external reward, and the asymptotic strength declines along the chain with increasing distance from the "output" classifier.

The formulae considered so far are such that, at each point in time, the bid is a function only of the strength of the classifier C_i which is activated. An interesting modification to the standard bucket-brigade algorithm (defined in Eq. (6.2)) consists in taking into account also the strength of the classifiers which have posted the message which C_i matches, thus introducing a further "memory" mechanism into the system. This can be done (Holland, 1986) by introducing another variable, the "relative support" V_i of a classifier, which measures, in a certain sense, the "value" of the messages which C_i matches.

Let us suppose then that C_i matches a certain number of messages, whose set will be represented by M. Let us call G the set of classifiers which posted, at in the preceding step, the messages which belong to M and let G' be the set of all the classifiers which posted messages at the preceding step. The relative support V_i and the bid b_i are then defined as follows:

$$V_i(t) = \frac{\dfrac{\sum_{j \in G} b_j(t-1)}{\Omega}}{\dfrac{\sum_{k \in G'} b_k(t-1)}{\Phi}} \tag{6.3}$$

$$b_i(t) = \alpha \gamma_i s_i(t) V_i(t) \tag{6.4}$$

Each classifier in G will receive a payment equal to b_i/Ω, where Ω and Φ are the cardinalities of the sets G and G', respectively. The time dependency has been explicitly shown to stress the fact that the bid at time t depends, through the support, upon the strengths at the preceding time step. The bid of a classifier which posted a given message is sometimes called the "intensity" of that message.

The reason why more specific classifiers are privileged in the competition stage is related to the development of "default hierarchies" for handling those rules which are often, but not always, valid. For example, a classifier system, employed in a task of animals identification, could develop a rule such as "if it lives in the sea, then it is a fish". When a dolphin is met, then the validity of the rule is contradicted. On the other hand, even though not always true, the rule is not useless, since it provides a starting hypothesis which is often valid. One possible solution consists in the introduction of another rule which specifies, for example, that "if it lives in the sea, and it surfaces to breathe, then it's a mammal".

The preceding rule, i.e., the more generic one, remains in the system, and is activated when one only knows that the animal lives in the sea. In the case where one knows that it surfaces to breathe, both rules match: the introduction of the specificity into the bid expression tends to favour the prevalence of the more specific rule, i.e., that which handles the exceptions.

It is also to be noted that this mechanism works only if the strengths of the specific and the generic rules are roughly equal. In practice, the usual rules (Eqs. (6.2) and (6.4)) show some drawbacks, already pointed out by Riolo (1987). Let us suppose that, at a certain time, we have two "chained" classifiers, i.e., such that one of them posts the message which activates the other, and that both of them are capable of beating the competition and of posting their own message. Let us also suppose that, at the time step considered, they both have about the same strength, but that the first activated classifier, G_g, is more generic than the other, C_s. Under these conditions, C_g receives a payment from C_s which is greater than the price which it has to pay, in its turn, to post its own message. If, however, it had been the most specific classifier C_s which had posted the message matched by C_g, then it would have been paid less than its own bid.

In any case, therefore, specific classifiers are at a disadvantage, given equal strengths, compared to generic ones, and this leads to an increase of the strengths of the latter.

Indeed, bucket brigade dynamics is regulated by the bids, and not by the strengths, and the possible evolution towards a stationary state tends to equalize the bids within a single chain. This, however, leads to the generic classifiers having a greater strength than the specific ones and, since the genetics is regulated by the strengths, the generic ones will have a greater rate of reproduction and will tend to predominate in successive generations.

In order to avoid this effect it is obviously possible to use Eq. (6.1), instead of Eq. (6.2). This would, however, preclude the possibility of implementing the default hierarchies for handling exceptions, and for privileging the more specific classifiers.

A more elegant solution is based upon the observation that, in the bucket brigade algorithm, the bid has two different uses: i) to choose the winners in the competition for posting and ii) to define the amount which the winners have to pay. It could therefore be useful to modify the algorithm in order to distinguish between these two aspects: the choice of the winners would be made according to Eq. (6.2), privileging the more specific ones, while the strengths would be adjusted according to Eq. (6.1), in order to avoid increases in the strengths of the generic classifiers. The proposed modification thus consists of the introduction of an additional variable p_i, which measures the entity of the price paid by the classifier C_i for posting its message. If b_i is the variable which regulates the competition, and p_i is the amount to be paid, then the equations become:

$$p_i = \alpha s_i$$
$$b_i = \gamma_i s_i \quad . \tag{6.5}$$

If the support is included, instead of Eq. (6.5) we obtain:

$$p_i = \alpha s_i V_i$$
$$b_i = \gamma_i s_i V_i \quad . \tag{6.6}$$

Riolo has introduced the two different bids in his implementation of classifier systems. The main difference between his formulae and Eqs.(6.5), (6.6) is that in the former case several parameters are to be defined, while in the latter case the formulae are quite straightforward: actually, Eqs. (6.5) and (6.6) correspond to particular choices of parameter values in Riolo's formulation. As we already noticed, classifier systems are rather complicated, with several adjustable parameters, and it is important to keep the possible sources of further complication at a minimum, in order to achieve a reasonable understanding of their properties.

As we have already seen, classifier systems are related to both production systems in classical AI and to dynamical systems. The relationship with the former is evident: the whole armoury of rules, pattern matching, conflict sets is very similar to that of production systems. The main difference lies in the fact that several rules can fire simultaneously, while in most AI systems the conflict resolution mechanism selects only one rule at a time. From this point of view classifier systems may be considered as a version of production systems with a greater degree of parallelism.

The relationship between classifier systems and dynamical systems lies in the association with each rule of the numerical variable which measures its strength. The bidding and rewarding mechanism described above leads to a dynamics which, over time, alters the values of the strengths. Take the case of a fixed population, that is without activating the genetic operators. Let us also choose the form of bidding and payment described by Eq. (6.5). At each step the strength of a classifier which has posted a message is then reduced by an amount equal to the payment p_i.

The other terms which can alter the strength come from payments by other classifiers and by the reward/punishment from the external environment. Let us also add an ever-present "dissipative" term, which can be interpreted, in the economical metaphor of the bucket brigade, as a small tax which each classifier has to pay at each step. This term, linear with respect to the strength, has the effect of depressing the strength of useless classifiers, which never enter into action. With the above hypotheses, the evolution equation for the strength s_i of the i-th classifier will then be:

$$s_i(t+1) = s_i(t)(1 - \beta) - \alpha s_i(t) F_i(t)$$

$$+ \sum_{j \in P_i(t)} \frac{\alpha s_j(t)}{n'_j(t)} + R_i(t) \tag{6.7}$$

where β is the coefficient of the dissipative term ($\beta \ll \alpha < 1$).

In this expression, $F_i(t)$ is a function which takes a value of 1 if the classifier i posts its message at time t, otherwise it takes the value 0. Analogously, $R_i(t)$ is equal to the external reward if the classifier i, at time t, posts an external message (i.e. a message which is readable by the effectors) and is otherwise 0. Finally, $P_i(t)$ is the (possibly empty) set of classifiers which have been able to post their messages at time t, exploiting a message posted by the i-th classifier at time $t - 1$. The denominator takes into account the fact that each classifier of P_i could have been put in a firing position not only by C_i, but by n'_j different classifiers.

It might seem that Eq. (6.7) is linear with respect to the strengths. However, this is not the case, since the various time-dependent functions are not simply forcing functions, externally assigned, but are functions which reflect the behaviour of the system. Suppose, for example, that a threshold mechanism is chosen to authorize the inclusion of messages in the message list. In this case, the function $F_i(t)$ can be expressed as the product of a function $F_i'(t)$, which takes a value of 1 if C_i matches a message and is otherwise 0, and of the Heaviside function $H(s_i(t) - \theta)$. This expression explicits the nonlinearity of Eq. (6.7). Also the choice of a competition mechanism based upon a message list of finite length would introduce a nonlinearity into the system.

Also, the set P_i would certainly be empty if C_i had not posted any messages at time $t - 1$. Thus, in reality, we are dealing with a system of higher than first order.

If we were to introduce the support into the evolution equations, we would obtain an explicit nonlinearity and a delay term, through a term which would multiply strengths at time t by strengths at time $t - 1$.

The system of equations obtained in this manner is too intricate to be described analytically. This does not mean, however, that it is impossible to obtain any indications about the behaviour of the variables and upon the role of the various parameters. In the next section we shall provide examples of this statement by discussing the results of studies upon the dynamical properties of classifier systems.

It is worth mentioning that the kind of classifier systems we have described so far are also called in the literature the "Michigan approach"; this is contrasted with the "Pittsburgh approach" (DeJong, 1988) which, however, shares several important features of the former. In both cases, genetic algorithms are applied to rule-based systems: the main difference lies in the fact that, while in the "Michigan approach" the individuals, upon which genetics operate, are identified with single production rules, in "Pittsburgh type" systems they are identified with entire production systems.

The genetic operators then operate by crossing and mutating entire rule systems, instead of individual rules. The evaluation mechanism also is defined in such a way as to estimate the fitness of entire production systems.

These systems (either of the "Michigan" or "Pittsburgh" type) are in an experimental stage and many properties still have to be understood. Some attempts have been made, however, to apply them to problems of a certain complexity, in order to test their possibilities. In any case, the "applications" documented so far regard simplified problems with respect to real-size problems.

The most well-known example is the control of a gas transport line (Goldberg, 1985). A classifier system, starting from zero a priori knowledge – i.e., from random rules – learns to decide upon the correct manoeuvres to be carried out in a compression plant, as a function of variables such as inlet and outlet flows, upstream and downstream pressures, temperature, etc. It also learns to detect leaks along the pipeline.

Other examples of application of genetic techniques to learning problems (Davis and Steenstrup, 1987; Goldberg and Holland, 1988) regard poker playing heuristics (Smith, 1985), multi-class pattern discrimination (Schaffer and Grefenstette, 1985) and strategies for the prisoner's dilemma (Axelrod, 1987).

As was mentioned above, there are also several applications of genetic algorithms to optimization problems, but this subject goes beyond the aims of the present work.

6.4 The Dynamics of Classifier Systems

This section will examine the dynamical behaviour of classifier systems. Particular attention will be given to the kind of rules which the system is able to discover, to the learning rates and to the stability of system performance under the action of genetic algorithms.

In order to address these issues, we will largely rely upon simulation results. The simulations have been carried out using the CFS-C package developed by Riolo at the University of Michigan.

In this system, every classifier has two conditions and, as was previously discussed, the presence of multiple conditions imposes a choice about how to deal with the pass-through mechanism ("#" in the action part). In the simulations discussed here, their positions are filled with the corresponding symbol of the first of the two conditions of the classifier or, if this is a "#", by the corresponding symbol in the message which matches the first condition.

CFS-C is a domain independent simulation package. In order to better understand classifier systems, we will now discuss in detail their behaviour in a specific case, proposed by Robertson and Riolo (1988), consisting of the learning of a sequence of letters like, for example, "nets". The sequence is assumed to be continuously fed to the system, so that the effective input to the system itself is a periodic sequence (such as "netsnets...").

The classifier system receives in input, through a suitable detector, the information about which is the "current" letter, i.e. the one read at time t, and its task consists in guessing which letter will follow at time $t + 1$. Note that, in order to analyze the system's learning capabilities, its input window is composed of one letter only. The letters are coded in binary form to allow use of the classifier system previously described.

In order to fix a convenient notation, let us consider the following classifier: $D : a, D : a/G : b$.

The two conditions are separated by a comma, while the action part is separated by a slash. Every condition and action has the same format as that of the messages, and is composed, in the shortened notation used here, by information concerning the "channel" in which the message has to be read or written (uppercase) and information about the letter (lowercase).

In the examples discussed here, there are four channels: one of them, marked with the letter D, is reserved for messages coming from the detectors, another (G) for messages for the effectors, and the other two, labelled with the letters X and Y are reserved for internal messages (which can be read or written only by other classifiers).

For example, if there is a message from the detectors which states that the current letter is "a", the classifier $D : a, D : a/G : b$ sends a message to the effectors with which it forecasts the next letter as "b".

In the classifiers and messages actually manipulated, the channels are coded with two binary digits, subject to mutations according to the usual genetics rules.

The letters are coded in a redundant manner, through the seven terminal bits of the message. The messages, in the simulations described here, are sixteen bits long: besides the nine bits used for coding channel and letter, the remaining seven are free, i.e., they are not interpreted by the detectors or by the effectors: they are therefore hidden variables, with no direct interaction with the external environment.

Note also that, because of the "don't care" and "pass through" symbols, a single condition or action may involve more channels and multiple letters. For example, the classifier $DG : a, X : b, c/G : d$ can post its message (guess "d") if, on the message list, there is one message, coming from the detectors (D) or from a previous guess (G), which says "a", and an internal message (channel X) which says either "b" or "c".

There exist two fundamentally different types of letter sequences, "unambiguous" and "ambiguous" ones.

In the case of an unambiguous, or Markov, sequence, knowledge of the current letter univocally determines the following one (like, e.g., in the word "neural"). The system has therefore to learn a series of heteroassociations of patterns, of the type $n \rightarrow e, e \rightarrow u, \ldots, l \rightarrow n$.

The situation is much more complex in the case of non-Markov, or ambiguous sequences (like e.g. "connectionist") where it is not possible to predict the next letter knowing only the current letter. In this case, the system must be able to build a self-organized "internal" memory, which allows it to make a correct forecast. In the case of a non-Markov sequence, therefore, the system must be able to generate, through self-organization, some kind of a representation of time. As is well-known, problems of this type are amongst the most difficult ones in AI.

Therefore, in predicting letter sequences, the simplest task consists of learning unambiguous sequences. The first problem to be addressed regards the possibility itself of learning the right rules, starting initially from a random set of classifiers.

Classifier systems, with cover detector and cover effector mechanisms, demonstrate the capability of learning this task, provided that suitable values are chosen for some critical variables. In particular, the average specificity of the classifiers is crucial: in fact, overly generic classifiers (with a relatively high "don't-care" frequency) will tend to be activated in too many different circumstances, leading sometimes to wrong guesses.

It has been experimentally observed that the presence of a fraction of "#" symbols of the order of 1/10 or less leads to reliable task learning. With large fractions of "don't care" learning is often very slow and the system is occasionally subject to instability with declining performance. In particular, this happens when the initial population is chosen with the same frequency of all three symbols $(1, 0, \#)$.

In the case considered here, the external input is periodic, and it tends to produce periodic trends in the classifier forces. Although an analytical treatment of the equations of classifier systems is, as we have seen, impossible, some particular cases can be analyzed and these can help us to understand what happens even in more complex situations by illustrating some typical behaviours.

Let us take a system without genetics, in order to analyze its bucket brigade dynamics. Assume that we have a periodic input with period T, and let us analyze the evolution equation of the strength of a classifier C (to simplify the notation, classifier subscripts will be omitted). Using a stroboscopic analysis (Minorsky, 1974), we will study the system with a time interval equal to the period of the external environment. From the equations of Sect. 6.3 we obtain:

$$s(t + T) - s(t) = -\left\{\sum \text{bids}\right\} + \left\{\sum \text{external reward}\right\}$$
$$+ \left\{\sum \text{payments from classifiers}\right\} - \text{decay} \tag{6.8}$$

All the summations refer to the interval $[t, t + T]$. The last term describes an exponential decay; however, if the decay constant is sufficiently slow with respect to the period of the environment ($\beta \ll 1/T$), it can be approximated with a linear term.

The other three terms depend upon the type of classifier. To illustrate the method, let us suppose that the classifier is of the type $D : a, D : a/G : b$, which makes a bid only once per cycle and provides a correct guess (like, for example, in a non-amiguous sequence containing the subsequence "ab"). The technique used here can be straightforwardly generalized to classifiers which are activated several times, which post internal messages, etc.

Given these assumptions, the bid is made only once and, by adopting Eq. (6.5), the corresponding payment is independent of both specificity and support, its value being $\alpha s(t')$, where t' is the time at which C bids, somewhere in the interval between t and $t + T$. We will assume that $t' = t$, that is, our analysis begins at the instant in which C finds a match. Note that the same formula is valid also in the case with relative support, as long as it is constant (as, for example, in the case where only classifier C is activated at the moment the bid is made).

A classifier of this type receives no payment from other classifiers, so that the second term in Eq. (6.8) is zero, while the third is equal to the reward from the external environment. We will assume that the reward is the same for every letter, and that it is divided amongst all classifiers which formulate the correct hypothesis. Since we have assumed the absence of genetics, the term representing payment from the external environment is constant (and equal to R).

We thus obtain, given the above assumptions, the linear equation

$$s(t + T) = s(t) - \alpha s(t) + R - \beta T s(t) \qquad (6.9)$$

The asymptotic solution of this equation, with period T, can be obtained by putting $s(t + T) = s(t)$, so that:

$$s = \frac{R}{\alpha + \beta T} \qquad (6.10)$$

The asymptotic state is periodic, and Eq. (6.10) provides an estimate of the maximum value, while at the other points in time a dissipative decay of the peak value is observed (Fig. 6.2).

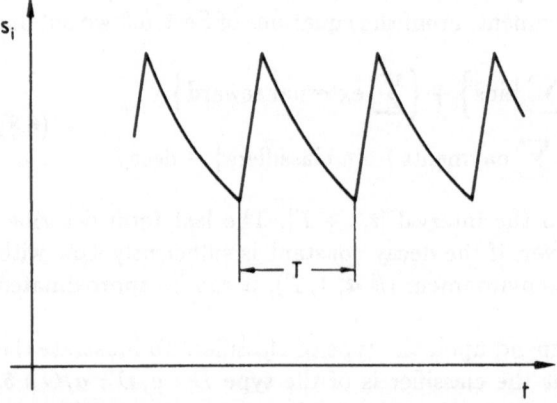

Fig. 6.2. Qualitative asymptotic behaviour of a classifier strength versus time in the learning of a Markov letter sequence (in a continuous time approximation). The input is periodic. In the interval between t and $t + T$ the classifier posts one external message which is rewarded

Various simulations, carried out under controlled conditions, have confirmed the validity of the approximate equation obtained in this manner, as well as some of the variants which describe other specific cases. Also, in analyzing the periodic solutions, it is easy to relax the linear decay approximation, leading to the exact formula (Compiani et al., 1988; Simonini, 1989):

$$s = \frac{R}{1 + \alpha(1 - \beta)^{T-1} - (1 - \beta)^T} \cdot \qquad (6.11)$$

Note that this expression is valid for the case where C can post once for each cycle. If there are several "synonym" classifiers which are activated by the same message from the detectors and which formulate the same prediction, the asymptotic value will be reduced in proportion to their number (in fact, the reward R will be divided by the total number of synonyms).

If the population becomes more numerous than the maximum length of the message list, each classifier will then, on average, fire less than once per cycle (in fact, the choice mechanism is a probabilistic one, with a probability distribution proportional to the strength). In this case, the previous expression still holds for the average strengths (when R is suitably re-normalized), but the behaviour of the strengths will lose its cyclic character, becoming a stochastic process due to the probabilistic choice mechanisms.

A simple application of the above analysis concerns the a priori introduction of correct rules with strengths which are too high. Assume that we start with a classifier population composed of a certain pool of random rules, and of several correct rules, for example, $D : a, D : a/G : b$. As we have seen, the bucket brigade tends towards asymptotic values, equal to at most $R/(\alpha + \beta T)$.

If the initial values of the strengths are higher, the strength of the "good" classifiers, which are immediately activated, will decrease towards the asymptotic value. The other classifiers, which are not activated, will decay more slowly, due only to the action of the dissipative term. If the genetics operates during this transient, it will substitute the weaker classifiers, i.e., the correct ones: in this way all a priori knowledge engineering previously carried out will be rapidly lost.

A simple technique for protecting the rules introduced a priori would be to block their attack by the genetics, identifying them as unmoveable rules, but this could turn out to be too rigid a solution which might inhibit further learning. Also the problem of the stability of the valid rules discovered by the genetics is crucial.

Let us observe that the "genotype" (which indicates, by analogy with the biological case, the microstructure of the classifiers, that is, the detailed sequence of the symbols 1, 0 and #) can not be invariant with respect to the genetics, except in very peculiar cases, like that of a completely uniform population under the action of crossover alone.

There is another interesting case where the genotype remains invariant: the case of a system in which only the triggered genetics is active, and which has reached an asymptotic state of the bucket brigade, corresponding to a success level of 100%.

Excluding very peculiar cases, the genotype will remain unchanged only if the "background" genetics (i.e., that which is not triggered) is not activated. Therefore, in the case where the genetics continues to operate indefinitely, fixed points of the genotype do not exist. In classifier systems there is a finite number of classifiers (M), each one of a finite length $(2n)$, defined from an alphabet (of three symbols): we are thus dealing with a finite number of different configurations. Therefore, the evolution of such a system will regenerate, after a recurrence time $\leq 3^{2nM}$, a population which had already been obtained. The upper bound of the recurrence time grows exponentially with the total number of symbols ($2nM$).

However, having a finite number of classifiers does not imply a periodic asymptotic behaviour. The strengths are, in fact, continuous variables. Moreover, the evolution of the system starting from an arbitrary state, as we have seen, is not deterministic, but is affected by randomness.

In order to evaluate the classifier systems's cognitive properties, what is important is to maintain a given performance level, rather than maintaining an invariant population. The genetic mechanisms proposed (substitution of only the weaker classifiers with mutants of the stronger ones) aim, in fact, at preserving system performance which could be compromised by a less than cautious use of genetics.

Generally, in the case of a non-ambiguous sequence, the solution found is of an intuitive type: the genetics succeeds in generating classifiers which perform straightforward associations between the current letter and the next one (e.g., $D:a, D:a/G:b$) reaching a success level of 100% (Fig. 6.3). This level typically lasts for al least tens of thousands of steps.

Occasionally, this performance level is destroyed by a perverse but interesting mechanism. The "good" classifiers, capable of making the correct associations, gradually proliferate; their descendants may not have the same genotype, but they have the same phenotype, in the sense that, under all conditions actually experienced, they behave exactly like their parents. Therefore, there are various "synonyms", for example, all those of the type $D:a, D:a/G:b$.

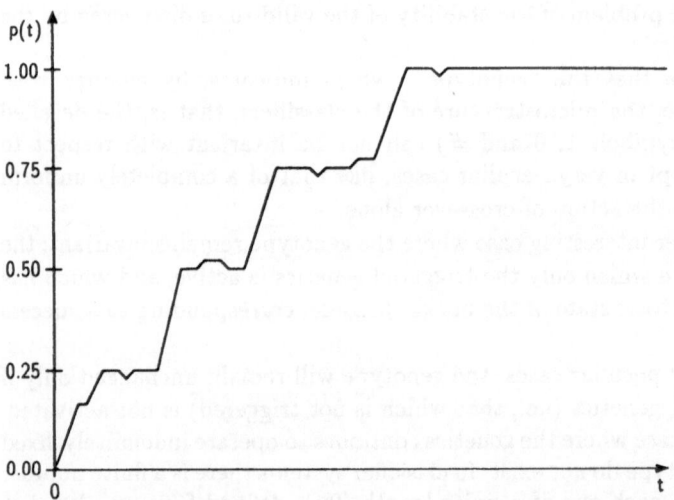

Fig. 6.3. Growth of performance ($p(t)$ is the fraction of successful guesses) as a function of time, for the markovian sequence "*abcd*"

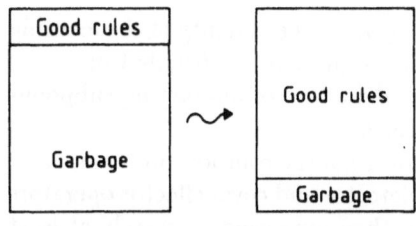

Fig. 6.4. A dangerous situation: subpopulations of good classifiers proliferate and overall strength equalization takes place

Slowly the list of classifiers fills up with subpopulations of the same phenotype (see Fig. 6.4): the subpopulation of those which read "*a*" and predict "*b*", etc. The subpopulations all have roughly the same numerosity and strength : the external reward is, in fact, divided amongst all the classifiers which post the right guess, and this inhibits the excessive proliferation of a subpopulation.

This situation can easily degenerate; the genetics, in fact, introduces new classifiers with a strength equal to the average strength, which in this case essentially coincides with the strength of all the classifiers, and a new-born classifier has a significant probability of prevailing – due to the effect of the probabilistic mechanisms – over its ascendants and to post its own message to the effectors. The system then makes a mistake; let us suppose, for example, that the error regards the attempt to guess the letter "*b*", given that the current letter is "*a*".

In this situation the cover effector operator comes into action, and this generates classifiers which respond to reading the letter "*a*", but with random actions. The classifier responsible for the first error is penalized and decays, but the new classifiers also have a strength which is equal to the average one, and join in the competition to post messages. In this manner, the system enters into a turbulent phase, characterized by inferior performance levels and by the proliferation of classifiers which respond to the letter "*a*".

Eventually, the system recovers and succeeds in making the right rule emerge but, since the strengths are equalized, the mechanisms described above may repeat themselves. To summarize, the phase in which the strengths are all roughly equal is a turbulent one, with inferior performance with respect to the preceding period of stability.

It is possible to avoid the emergence of this destabilizing mechanism by changing the selection rule in the choice of winners in the competition to post messages. The phenomenon just described is, in fact, observed in the case where the size of the message list is small and the probability that a classifier which matches will be selected to post a message is proportional to its bid. This distribution only slightly privileges the high bidders compared to those with a lower bid, and this is particularly evident in the case where all the strengths are roughly equal.

By choosing a sharper probability distribution, however, a qualitatively different behaviour is observed: now, strength equalization does not take place and every subpopulation consists of a number of strong classifiers and of other synonyms which are much weaker and undergo irreversible decay, since their chance of posting during their lifetime is very low. The stability of the learning capacity, therefore, strongly depends upon the probability distribution.

A thorough quantitative analysis of the dynamics of interacting subpopulations can be found in Compiani et al. (1989c).

It should also be observed that the discovery of the solutions in the nonambiguous case is very slow, also when cover detector and cover effector operators are active. The discovery is actually slower than pure random search, at least if an external payoff is used which does not take into account the overlap of the actual guess with the correct one, but which simply rewards correct guesses and punishes all the wrong ones in the same way. The reason why the discovery is so slow is that, even if the right classifier is discovered, the mechanisms of competition and bidding may prevent it from firing, thus leading to its decay and to its loss. However, in the more complex and more interesting case of prediction of ambiguous sequences, classifier systems clearly outperform random search. They actually prove capable of developing an internal memory by self-organization.

Because of the greater difficulty of this task, compared to the previous one, it has not, so far, been possible to consistently obtain success rates of 100%, stable over tens of thousands of steps.

The analysis of the ambiguous sequence is interesting, not only because it tests the learning ability of the system, but above all because it shows several solutions which mix the inferential (pattern matching) and dynamical (competition based upon the strengths) features in an unexpected manner. A strong interaction is thus observed between the logical and the dynamical side of the classifier approach to cognitive systems.

Take, for example, the case where the sequence is "*abacabac...*": the forecasting of "*a*" is analogous to the preceding case, but the forecasting of the other two letters requires the creation of an internal message which remembers the letter read at the preceding instant in time. An intuitive solution (an "anthropomorphous" one, since it is the one which would be proposed by a knowledge engineer), which is capable of carrying out the task correctly, is represented by the following six classifiers:

1) $D : b, D : b/G : a$
2) $D : b, D : b/X : b$
3) $D : c, D : c/G : a$
4) $D : c, D : c/X : c$
5) $D : a, X : b/G : c$
6) $D : a, X : c/G : b$

The last two classifiers forecast the ambiguous portion, each using an internal message which represents the letter read at time $t - 1$, and a message from the detectors which describes the letter read at time t.

Surprisingly, numerous simulations show the development of a different representation, based upon the presence of at least one classifier which is not always appropriate, like for example $D : a, D : a/G : b$, which is capable of forecasting correctly only once every two guesses. In these cases, however, mechanisms emerge which ensure that these classifiers win the competition for bidding only when required ("dynamical" solutions).

An example of this kind is the following:

1) $D : b, D : b/G : a$
2) $D : b, D : b/X : b$
3) $D : c, D : c/G : a$
4) $D : a, D : a/G : b$
5) $D : a, X : b/G : c$

In the case where the message from the detectors is "$D : a$", and there is an internal message "$X : b$", the last two classifiers both match, and post their messages. There is thus a conflict between two different guesses: the conflict resolution mechanism, having to decide upon the output, chooses the one with the higher bid with probability one.

Therefore the mechanism works correctly if the bid of the last classifier is greater than that of the penultimate one. This occurs quite readily if a bidding with support mechanism is introduced, since the support of the last classifier is the sum of the intensities of the message coming from the detectors and of the internal one, whereas the support of the fourth classifier is only equal to the intensity of the message from the detectors: with equal strengths, the latter is readily overcome by the former.

One particularly interesting solution was observed in the case of the ambiguous sequence "aab": the classifiers in play were as follows (the values of the strengths are given in brackets):

1) $D : a, D : a/G : a$ (s_1)
2) $DG : a, D : a/G : b$ (s_2)
3) $D : b, D : b/G : a$ (s_3)

This solution was found very quickly, and it is stable if only the error-triggered genetics is present (see Fig. 6.5).

The third classifier solves the non-ambiguous part of the task, while the first two respond to "$D : a$" messages. The only difference between their conditions is that the second message can also use an internal message $(G : a)$ besides the one from the detectors. Note, in fact, that "G" channel messages can be read not only by the effectors, but also by other classifiers.

Let us then assume that at time $t-1$ a "b" was read: at time t an "a" is read, and the first two classifiers both match the message from the detectors. Using a message list with more than one place, both of them post their own messages: two contradictory messages then reach the effectors. The conflict resolution method is such that, under these conditions, the one with the highest bid prevails. Since $s_1 > s_2(s_1 \approx 3s_2$ in the simulation shown in Fig. 6.5) and both

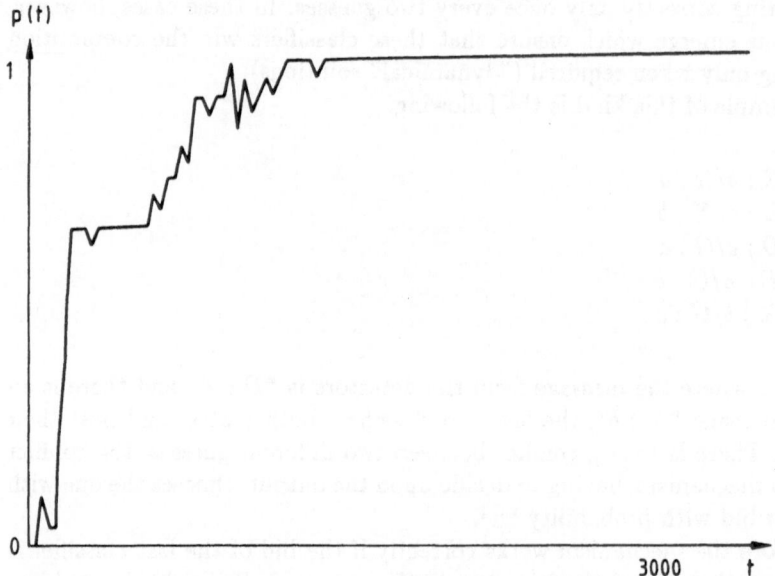

Fig. 6.5. Growth of performance $p(t)$ as a function of time, for the ambiguous sequence "aab"

the classifiers have the same support, the first one prevails, and the correct guess "a" is made.

At time $t + 1$, there are two messages in the message list, "$D : a$" and "$G : a$", the latter having been posted by the first classifier at the preceding time step. Once again, both (1) and (2) match, setting up their messages, but this time it is the second classifier (which matches both a message from the detectors and an internal one) which prevails in the conflict resolution and sets up the correct message $(G : b)$.

This solution may be considered as a form of "symbiosis" between classifiers (1) and (2), since the second, which is weaker and therefore gives way to the first one when the support is equal, takes advantage precisely of the high strength of the first one, and therefore of the high intensity of the message which it posted, to prevail at the right moment. In the absence of their partner, both (1) and (2) would formulate wrong forecasts in half of the cases, and would consequently be punished. The competition which develops, however, strengthens both of them, generating an exact prediction mechanism.

This solution resembles an oscillator which spontaneously develops in the system, synchronizing itself with the frequency of the signal from the external environment.

In the "anthropomorphous" solution, in a certain sense, the strengths are no longer essential once the learning stage is complete: in fact, one could suppress the dynamical part (strengths, supports, competition) leaving only the match mechanism active, without deteriorations in performance. The "dynamical" solutions, on the other hand, directly use the values of the bids, and essentially couple the pattern matching with the variations in the latter: this is particularly evident in the case of the solution with "symbiosis".

Another interesting solution observed relies on the use of hidden variables. Let us recall that a part of the string does not have any meaning imposed by the external environment. The system, in some cases, develops a method for coding the various situations based upon the use of these variables. Let $G : a\gamma$ and $G : a\beta$ be two messages on channel G, which both guess the letter "a" but differ in the silent part of the string. A sequence of classifiers for the prediction of "$abac$" is therefore:

1) $D : a, G : a\gamma/G : b$
2) $D : b, D : b/G : a\beta$
3) $D : a, G : a\beta/G : c$
4) $D : c, D : c/G : a\gamma$

One reason why, in some cases, it is not possible to consistently reach a 100% level of correct predictions is related to interesting kinetic considerations. Let us once again take the sequence "$abac$": as may be expected, the classifiers which correctly solve the non-ambiguous part of the sequence ("$G : a$") develop before the chained ones which succeed in distinguishing the non-ambiguity of the other part of the string.

The population, at a certain point in time, will then be composed of classifiers which correctly forecast the non-ambiguous part and, due to the effect of the cover operator, of classifiers which respond to the conditions "$D : a, D : a$" in an inappropriate manner. For example, we could have:

1) $D : a, D : a/G : h$
2) $D : b, D : b/G : a$

Classifier (1) acts at time t, and makes a wrong guess. Classifier (2) acts at time $t + 1$, makes the right prediction and is rewarded. We thus have (1) active at time t, while (2) is active and profitable at $t + 1$. These conditions are suitable for the triggered chaining operator, which may generate:

3) $D : a, D : a/X : a$
4) $D : b, X : a/G : a$

(4) carries out the same task as (2) but tends to prevail because of its higher support level: the system generates a chained solution also for the non-ambiguous part. If the chains proliferate, there will be a large number of classifiers of type (4), which will tend to fill the message list with their correct guesses. Under these conditions, it may happen that the genetics generates the necessary classifier

5) $D : b, D : b/Xb$

but it will have a very little probability of posting its own message, having to compete with the subpopulation of the synonyms of (4), and will decay.

A way to limit this effect consists in preventing the proliferation of subpopulations of synonyms; however, only genotypes can be directly checked by the CFS-C system, and one therefore resorts to limiting the maximum number of copies of identical classifiers, which has an indirect effect upon the proliferation of synonyms. Provided that such an indirect control is achieved, the performance level sharply depends upon the size of the message list (Fig. 6.6), since allowing more messages gives a higher chance to classifiers of type (5) to post.

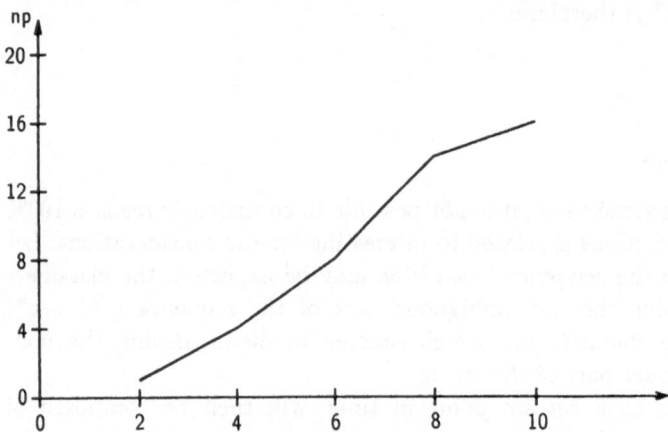

Fig. 6.6. Total number (np) of completely successful trials as a function of the message list size M. Twenty trials were made for every value of M. Total number of classifiers is 50

6.5 Classifier Systems and Neural Networks

We will now compare classifier systems with the neural networks discussed in the previous chapters. This study is motivated by the attempt to understand the relationship between these two classes of systems and by the hope that the differences suggest possible modifications and extensions of them. The approach which will be adopted for making this comparison consists in starting from a classifier system and constructing an equivalent dynamical network, which will be called the "image" network. In this process both the main similarities and the significant differences between the two approaches will emerge.

As we know, there are various versions of classifier systems and, in order to present the ideas, we will begin with a particular case and then generalize to more complex cases.

Let us take the case of classifier systems with a single condition (and a single action) and an evolution law without support (i.e., as in Eq. (6.5)). Initially, we will also assume that the classifiers do not present the symbol "#", being generated only from the alphabet $\{0, 1\}$. The competition rule remains to be defined for establishing which classifiers post their own message. We will first analyze the case where the competition is ruled by a fixed threshold mechanism, i.e. a classifier posts its message if its bid exceeds a fixed value θ.

At first sight, classifiers and neural networks seem to have almost nothing at all in common: in one case there are rules, messages, strengths and message lists, whereas in the other we have formal neurons, transfer functions, connections. However, let us try to associate every possible message with a neuron in a network: since there are 2^n possible messages, the network has the same number of nodes and, for reasonable values of n, it is therefore very large.

The analogous of the firing of a neuron, in this picture, should be the posting of a message. In other words, the nodes of the image network will be active, at a given instant, if the message which they represent is in the message list, and otherwise inactive. We thus have a boolean network.

A fundamental aspect of neural networks are the connections: since the nodes are associated with the messages, the connections are associated with whatever links the messages, that is, the classifiers. We may then suggest, for comparing neural networks and classifier systems, the following correspondence (Compiani et al., 1988; Miller and Forrest, 1989):

messages $\leftrightarrow$ formal neurons
presence in message list $\leftrightarrow$ firing
classifiers $\leftrightarrow$ connections

By adopting this point of view, we immediately see an important similarity between classifier systems and neural networks: the classifiers have an associated numerical variable (the strength) just like the synaptic connections, and in both systems learning occurs by operating upon the connections.

To be precise, let us represent the nodes of the image network as points in an n-dimensional space: in particular, as the vertices of a hypercube of unit edge which has a corner at the origin and all the others in the sector of the non-negative coordinates. In this way it is possible to associate each corner with a message in a natural way: the coordinates of the corner constitute an n-tuple of symbols taken from the alphabet $\{0, 1\}$ with which we can associate the message described by the same string. The connections (classifiers) thus join pairs of corners.

Let us first examine the activation rule: as we know, the message m_i is posted (that is, the i-th formal neuron of the image network goes "high", $x_i = 1$) if, amongst all the classifiers C_{ij} which have m_i as their action, there is at least one which i) matches with a message which was active at the preceding time step and ii) makes a bid which is sufficient to win the contest, i.e., which exceeds the threshold. Let us take $\{0, 1\}$ as possible activation values for the elements of the network, and let us call s_{ij} and b_{ij} the strength and the bid of the classifier C_{ij} whose condition and action are, respectively,

m_j and m_i. The activation rule is therefore:

$$x_i(t+1) = 1 \text{ if } j \text{ exists such that } b_{ij}(t)x_j(t) > \theta$$
$$x_i(t+1) = 0 \text{ otherwise}$$

This formula is similar, though not identical, to the Hopfield evolution law. The difference is that, in the latter case, the threshold is compared with the sum of the input terms (synapse * excitation products), while in the classifier case every input is separately compared with the threshold. The excitation mechanism is, in this case, less collective than in neural networks. This suggests a possible modification to classifier systems.

Let us now come to the learning algorithms. The bucket brigade alters the numerical value of the synaptic connections, thus playing an analogous role to that of algorithms commonly used in neural networks, such as back propagation.

The genetics, however, operates in a different manner, generating new connections which did not exist previously, and eliminating those of low intensity. There is no analogous mechanism in the neural networks which we have examined so far.

The idea which we thus obtain of the "image network" is that of a system with an enormous number of elements, of which, at any particular moment, only a small fraction is active. Of the vast number of possible connections, initially only a few are present. The network is operated with a supervised learning mechanism, which determines the values of the connection strengths.

When the genetics is activated, some connections are eliminated, and new ones are introduced. This allows the exploration of a new portion of the network, and the procedure is repeated. Also in neural networks with a high number of elements and incomplete connectivity, the idea of generating and testing new connections is interesting and could produce important results.

The point mutation operator, starting from a connection between two corners which will be called A and B, introduces a new one, such as, for example, $A' - B$, characterized by the fact that A' is very close to A. In this way we have an exploration in the neighbourhood of existing connections. The crossover operator, on the other hand, can generate also connections between points which are quite distant from the starting points.

The case analyzed so far refers to a particularly simple classifier system. Let us now examine some possible variants, and study the way they change the "image network".

First of all, take the case of classifiers with multiple conditions. Activation of the message corresponding to the action of classifier C therefore requires that all the messages corresponding to its h conditions be in the message list and that C's bid be greater than the threshold.

C will thus be represented in the image network not by one, but by h connections, all of the same intensity. If we call I^* the set of h active messages which match all the conditions of C, the activation rule will now be:

$$x_i(t+1) = 1 \quad \text{if} \quad I^* \text{ exists such that} \quad b_{ij}(t)\prod_{j \in I^*} x_j(t) > \theta$$
$$x_i(t+1) = 0 \quad \text{otherwise}.$$

This rule is similar to the evolution law of the so-called "sigma-pi units" (Rumelhart and McClelland, 1986), although the activation mechanism of classifier systems is of a less collective type.

Another variant, more frequently adopted in classifier systems, consists of using a competition mechanism which is not linked to the comparison between the bid and the threshold, but is based upon a limitation of the maximum number of messages which can be posted. If the candidates (i.e., the messages generated by the classifiers which find a match) exceed the maximum allowed number, the highest bidders are selected. A technique for managing this competition is based upon the presence of an "agent" – which could even be a second neural network – superimposed over the image network, which calculates the number of active elements and switches off the excess fraction.

Let us now examine the modifications to the image network required by the introduction of the matching mechanism based upon the use of "don't care" and "pass through" symbols. A classifier with "#"'s not only connects two messages ("neurons"), but two classes of messages. This implies that the learning process is capable of exploring sets of connections, instead of single connections, with a positive effect upon the rate of exploration of the possible solution space. It should be noted, however, that in this way different connections are correlated, while in neural networks the various connections evolve independently.

Finally, yet another variant, with respect to the simple classifier system described above, consists of introducing the classifier support. As will be recalled, this amounts to introducing a new nonlinearity in the classifier system equations. The effect of this change upon the image network is emphasized by the fact that the support of a classifier C is a continuous variable, equal to the sum of the bids of the classifiers which have posted the messages which activated C.

An adequate schematization of this case is based upon the use of a image network with continuous variables. The output of a neuron is thus a real valued variable, equal to the intensity of the corresponding message if it is on the message list, and otherwise zero. The intensity of a message is, by definition, equal to the bid of the classifier which posted it. In formulae, indicating with I the input and with O the output of a node (message), we will have:

$$I_i(t) = b_{ij}(t-1)O_j(t-1)$$
$$O_i(t) = I_i(t)H(I_i(t) - \theta)$$

where H is the Heaviside function. Therefore, the transfer function is, in this case, a linear threshold function.

In conclusion, classifier systems and neural networks both make use of the properties of dynamical systems, and of the changes in the "connections" between nodes to perform learning. The former exploit genetic mechanisms for exploring different variants of the system, while neural networks show a more collective activation mechanism, in which the activation of a node is more often the result of a sum of small contributions, rather than the effect of a single shot. A more detailed description of the relationship between classifier systems and neural networks, which has only been briefly outlined here, can be found in Compiani et al. (1988a). A similar approach is discussed in Miller and Forrest (1989), where the relationship between classifier systems and random boolean networks is analyzed.

Appendix 6.1 Implicit Parallelism

The discussion below relies heavily upon the presentation of implicit parallelism given by Schaffer (1987). We recall that every classifier C_i is a string of $2n$ symbols, generated from the alphabet $\{0, 1, \#\}$, and that a schema may be represented by introducing a "don't care" symbol (∗, not to be confused with "#" which is used to define a match).

Let $P(u, t)$ be the probability that a given classifier, at time t, is an instance of the schema u (i.e., that it belongs to the subset of the classifiers which contain u). Operating with finite populations, $P(u, t)$ is estimated from the relative frequency with which instances of the schema u appear in the classifier population at time t, $C(t)$. Let U be the hyperplane corresponding to the schema u, that is, the set of all possible classifiers which contain the schema u, and $U(t)$ the subset of U made up of those classifiers which are effectively present in the system at time t:

$$U(t) \equiv U \cap C(t) \quad .$$

The length of the schema, $l'(u)$, is defined as the number of positions between the first and the last meaningful symbols (i.e, other than "∗") in the definition of the schema. It is more convenient, however, to use the normalized length

$$l(u) = \frac{l'(u)}{2n - 1} \quad . \tag{A6.1}$$

Let us now assume that the genetics is activated periodically, every T steps. For simplicity, we will only use the crossover operator.

During the interval $[t, t + T]$ the classifiers belonging to $C(t)$ will have reached certain strength values, which are a measure of their "overall utility" for the system. The value of a rule strength, normalized with respect to the average strength, is a measure of the fitness of the corresponding classifier. An estimate of the fitness of a schema will be given by the ratio between the average strength of the classifiers which contain it and the overall average strength.

In formulae, the fitness ratio of a schema u is given by:

$$\Phi(u,t) = \frac{\frac{1}{m}\sum_{j\in U(t)} s(j,t)}{\frac{1}{M}\sum_{k\in C(t)} s(k,t)} \quad , \tag{A6.2}$$

where m is the cardinality of the set $U(t)$, while M is the total number of classifiers; $s(k,t)$ represents the strength of the k-th classifier at time t.

We still have to define the choice of the exemplars to be selected as "parents". In order to privilege the strongest classifiers, we will choose parents with a probability distribution which is proportional to the average strength; the probability $f(j,t)$ that classifier j will be selected as a parent will then be

$$f(j,t) = \frac{s(j,t)}{\sum_{k\in C(t)} s(k,t)} \tag{A6.3}$$

The Holland theorem on implicit parallelism provides a lower bound to the probability $P(u,t+T)$ of finding instances of schema u in the new population.

Let us first consider the case where the new population is composed only of copies of the preceding classifiers, generated as follows:

- a classifier is selected from $C(t)$ with a probability $f(j,t)$ given by the previous formula, and is copied into the new population $C(t+T)$;
- the same procedure is repeated M times.

Under these conditions, we can estimate the fraction of new classifiers which are instances of the schema u, that is $P(u,t+T)$. It will be equal to the fraction of instances of the schema in the initial population times the relative fitness of the schema:

$$P(u,t+T) = \Phi(u,t)P(u,t) \quad . \tag{A6.4}$$

The validity of this formula can be confirmed by its expansion; omitting the time variable for brevity, we have:

$$\Phi(u)P(u) = \frac{\frac{1}{m}\sum_{j\in U(t)}}{\frac{1}{M}\sum_{k\in C(t)} s(k)} \frac{m}{M} = \frac{\sum_{j\in U(t)} s(j)}{\sum_{k\in C(t)} s(k)} \quad . \tag{A6.5}$$

The r.h.s. is precisely the sum of the strengths of the classifiers in $U(t)$, divided by the total sum of the strengths, and is therefore equal, given the assumptions made, to the fraction of new-born classifiers which are "children" of classifiers in $U(t)$. This formula also demonstrates that $\Phi(u)P(u) \leq 1$, a necessary condition for its interpretation as a probability.

The formula thus obtained leads to the conclusion that the fraction of versions of a schema with greater than average fitness grows in successive generations. Repeated application of Eq. (A6.4) indicates that, if the average fitness of the schema remains approximately constant throughout various generations, then it reproduces in an exponential manner:

$$P(u, t + hT) \approx P(u, t)\Phi(u)^h \quad . \tag{A6.6}$$

Note, however, that the growth can be exponential only for a few steps; using the assumptions made, in fact, the fitness ratio of a schema is reduced over time, precisely because an increasing number of versions of that schema will be found in successive generations, thus increasing the average strength.

Let us now consider the most interesting case where the reproduction is accompanied with variations. This case is more complex, and it will be possible to define only an inequality defining the lower bound for $P(u, t+T)$. Although it would be possible to formally write a Master equation for the evolution of $P(u, t)$, it would be of such an intricate form as to be of no practical use.

There are several versions of this theorem on "implict parallelism", differing in their details. The formula given here is valid under the following assumptions:

- the population $C(t)$ is completely substituted when the genetics is activated, and the overall number M of classifiers is constant;
- the genetics is activated at time $t + T$, and is based only upon the use of the crossover operator. The parents are both chosen with a probability distribution proportional to the strength (Eq. (A6.3)).

Under these conditions, the Holland formula is:

$$P(u, t + T) \geq \Phi(u, t)P(u, t)[1 - l(u)(1 - \Phi(u, t)P(u, t))] \quad . \tag{A6.7}$$

The interpretation of the theorem is quite straightforward. Given the assumptions made, $\Phi(u, t)P(u, t)$ represents the probability that a classifier in $C(t+T)$ is a descendant of a classifier in $C(t)$. This has already been demonstrated in the previous example, relating to reproduction without variations.

The term in the square brackets is the expected fraction of those classifiers which are in $U(t)$ and which will remain in U after the application of the genetics: its value is equal to 1 minus the probability that the genetics destroys its belonging to U. This, in its turn, is equal to the product of two independent events: i) that the cross point falls within the interval of values which define the schema ($l(u)$) and ii) that the partner in the crossover is not also a member of $U(t)(1 - \Phi(u, t)P(u, t))$. The sign $\leq$ is motivated by the fact that other classifiers in $U(t + T)$ can be generated by parents which were not in $U(t)$.

Take, for example, the limiting case, corresponding to the minimum probability of "schema destruction", that is a one-symbol schema ($l(u) = 0$). The theorem then states that:

$$P(u, t + T) \geq \Phi(u, t)P(u, t) \quad .$$

This expression guarantees a rapid selection of the schemata whose performances are markedly superior to the average ones. The worst case, from the point of view of the selection, is that of a schema of maximum length: $l(u) = 1$. In this case, we obtain

$$P(u, t + T) \geq [\Phi(u, t)P(u, t)]^2 \ .$$

The effect here is that of more severely punishing the classifiers whose performances are below the average. It is also true, however, that even classifiers with above average performances can be destroyed by crossover if they are not sufficiently represented in the population $C(t)$. A sufficient condition in order that $P(u, t + T) > P(u, t)$ is that

$$P(u, t)\Phi(u, t)^2 \geq 1$$

holds. For this to be satisfied, it is therefore required that

$$P(u, t) \geq \Phi(u, t)^{-2} \ .$$

For example, a schema with twice the average fitness will see its own relative frequency increase if it is already present in at least a quarter of the classifiers.

In conclusion, crossover tends, above all, to select short schemata with above average performance.

7. Problems and Prospects

7.1 Introduction

In the last three chapters we have examined various dynamical systems for artificial intelligence, illustrating some of the characteristics of the dynamical approach outlined in Chap. 2. There were, however, some questions, both of principle and of a technical nature, which could not be faced until the various specific models had been examined, and which will now be discussed.

One of these questions regards the role of knowledge representation in neural networks and, in particular, the definition of the semantics of the single nodes of a neural network. In other terms, for an external observer, what is the meaning of the activation of a specific neuron in the network?

This problem will lead us to consider, in Sect. 7.2, the relationship between local and distributed representations, and to analyze the meaning of the so-called "sub-symbolic approach".

Another question which carries both a conceptual and practical importance regards the essential nature of dynamics: considering, for example, the symmetrical Hopfield model, one immediately realizes that it works as an associative memory, associating an initial state X_0 with the corresponding attractor X_∞.

It would, however, be possible to obtain the same type of association without making use of any explicit dynamics, for example by using geometrical techniques or projection operators.

It is therefore important, in particular for those interested in building artificial systems, to compare the use of dynamical systems with that of projection techniques. This topic is addressed in Sect. 7.3, where also the role of chaotic dynamics and of fractal attractors in neural networks is mentioned.

So far, in this volume we have stressed the merits of the dynamical systems approach to artificial intelligence compared to the classical approach. As we have seen, dynamical systems can readily provide properties which are still lacking, or at best are very difficult to achieve with logic based systems.

However, the artificial simulation of cognitive processes is a formidable problem and, at least so far, no single line of attack has been capable of providing convincing results from every point of view.

In particular, the dynamical approach also presents significant limits and disadvantages, which are analyzed in Sect. 7.4. This detailed examination leads us to the conclusion that the artificial simulation of high level cognitive processes requires the introduction of systems which integrate both the dynamical and the classical approaches, combining the strengths, and not the weaknesses of the two paradigms. Although the results in this direction are still quite limited, possible lines of development and their perspectives are discussed.

7.2 Knowledge Representation

Let us now approach the important problem of the semantics of node activation in a neural network, that is the meaning which can be attributed to the state of individual neurons, or of groups of neurons.

For example, in Chaps. 2 and 4 we considered the recognition of letters using Hopfield networks (Fig. 2.1, reproduced here as Fig. 7.1). In this case, all the neurons in the network are visible units, and their meaning, which is completely defined by the "external environment", is that of the pixels of an image.

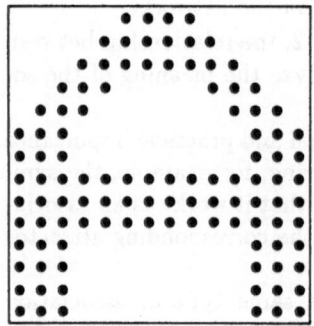

Fig. 7.1. A pattern can be associated to a dynamical system. In this case, the letter "A" is represented by boolean variables, associating a value "1" to black points and a value "0" to white points

The pattern which has to be recognized (the letter) is defined by the co-activation of several neurons, i.e., it possesses a representation distributed over a certain number of elementary units.

In this case, the attribution of a meaning to the individual neurons is quite straightforward. This particular representation has several disadvantages, such as the fact that it is invariant neither for translation nor for rotation. It is, however, possible to use neural networks with representations of a different kind. Remaining within the field of image recognition, it is known, for example, that it is often convenient to extract several features from the image, and to classify the latter on the basis of the formers.

It is possible to use these variables as inputs to a neural network which then learns to recognize the various letters according to the usual learning algorithms. The activation of a neuron in the input layer of this network could therefore represent, for instance, the presence of vertical lines, or the value of one of the moments of the probability distribution. In this case, however, something has changed with respect to direct recognition, since we have introduced a pre-processing system, according to the following scheme:

INPUT → PREPROCESSING → FEATURES → NEURAL NETWORK

The example of Fig. 7.1 may be considered a particular case of this scheme, where the extracted features are simply the values "black" or "white" to be associated with each pixel.

It is clear, however, that the nodes of a neural network can not only be associated with pixels or features of an image, but they can represent arbitrary types of information, including different kinds of signals (e.g., speech, radar, sonar) but also "symbolic" information, like e.g. the symptoms of a diagnosis system.

Let us consider the following example, which recalls one of McClelland, Rumelhart and Hinton (1986). Suppose that we want to learn to recognize our surroundings, and that there is information in input about the presence of certain objects. Table 7.1 shows the case of four different surroundings (a bathroom, a kitchen, a city and a countryside) each of them being characterized by the presence of certain features.

Table 7.1

Training set

Pattern 1: Door, window, bathtub, shower, soap, razor
Pattern 2: Door, window, oven, refrigerator, dishes, cups
Pattern 3: Open air, sky, traffic lights, cars, streets, shops
Pattern 4: Open air, sky, grass, flowers, trees, butterflies

Examples of recognition

Initial state	Final state
Door	Door, window
Door, sky	Null state
Bathtub	Door, window, bathtub, shower, soap, razor
Shower, oven	Door, window
Bathtub, traffic light	Null state

The training set is composed of four typical examples described in the first part of the table; the second part gives some examples of correct classification (Serra et al., 1988) obtained by using the asymmetric rule introduced in Sect. 4.4 for a boolean network (Eq. (4.38)). The nodes of the network are now associated with the various features which may be present, and the activation of a node corresponds to the presence of the corresponding feature.

Also in this case a distributed representation is adopted, so that the concept of "bathroom" is not represented by a dedicated node, but by the co-activation of several *microfeatures*: door, window, bathtub, shower, soap, razor. Analogously, the concept of "room" is represented by the characteristics common to the rooms considered: "door" and "window".

It can be seen here how this network can handle the correct level of classification, even in the presence of two contrasting pieces of information: if the information is "there's a shower" and "there's an oven", the network excites the aspects common to all the indoor environments (door and window), whereas if the information is "there's a bathtub" and "there's a traffic light", no nodes are activated.

An interesting utilization of neural networks with nodes which represent symbolic concepts is the understanding of natural language (Feldman, 1988). Let us consider, in particular, the interpretation of ambiguous sentences, such as, for example, the statement "the astronomer has married a star" (Waltz, quoted in Fogelman, 1985).

In this case, one can use a dynamical network whose nodes represent the various possible alternative interpretations of the terms used. Nodes which represent mutually incompatible interpretations will have inhibitory connections (e.g., "to wed" and "celestial body"), while those corresponding to congruent ones will have positive connections (e.g., "to wed" and "movie star").

The input sentence provides a certain initial activation of the various nodes which correspond to the possible interpretations: this activation is not the same for every interpretation but reflects the most frequent use of the term. For instance, "star" will more strongly activate "celestial body" than "movie star".

The network evolves starting from the initial state, through one of the usual updating algorithms based upon the sum of the inputs coming from the connected nodes, until it reaches an asymptotic state, which hopefully corresponds to a meaningful interpretation of the statement, by giving higher activation values to mutually consistent interpretations of the terms.

In this example, one can see how the initial, inconsistent interpretation ("a scholar of astronomy weds an astral body") evolves towards the self-consistent "a human being of male sex weds a movie star".

So far, we have considered systems without any hidden units, where the meaning of the single neurons is prescribed externally (i.e., by the experimenter), and where the different classes (patterns) are not prescribed, but emerge spontaneously during the learning phase in a distributed manner, and correspond to the simultaneous activation of a number of neurons.

Such a coding of the patterns presents some advantages: it is not necessary to pre-define the classes, and recognition is robust and fault tolerant. However, in a complex information processing system, there exist some points where information has to be abstracted and concentrated.

Let us consider, for example, a word recognition task, where recognition should be independent of the fonts used for printing the letters. All the different possible fonts which indicate a given letter must, at a certain stage, lead to the same situation. Let us suppose, for example, that a hypothetical word recognition network has been taught the word "ABHORIGENES" (uppercase letters). Such network would then be able to automatically correct a word such as "ABHORIGEMES" to "ABHORIGENES", but it should also be capable of doing the same with "abhorigemes".

If the word learning and recognition algorithm were to operate directly upon the distributed representation of the single letters, it would be necessary to form the suitable associations between all the possible fonts.

This is not only seriously inefficient, but poses problems of co-adaptation of different connections: for example, the learning of "ABHORIGENES" reinforces connections between the neurons which realize distributed representations of H in the third positions and those which realise O in the fourth. But how does the corresponding reinforcement of the connections for "h" and "o" come about? It would clearly be unsatisfactory if it required a separate learning.

Therefore, the treatment of information even in a moderately complex system requires that the information relating to input patterns be coded in an abstract manner, in order to allow hierarchical information processing.

A simple way for achieving such an abstract coding is the use of "coding fields" (Dreyfus, 1986), which is based on the idea of enlarging the original network by adding a new part, which codes higher level information (Fig. 7.2).

 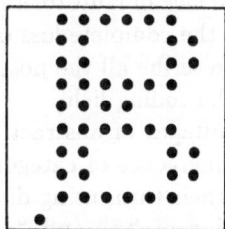

Fig. 7.2. Coding Fields. The coding field is here the bottom row of the network: its first element codes for "A", the second for "B", etc.

For the sake of definiteness, let us once again take the case of character recognition, and suppose that in the training set several versions of the same letter are present, for example several variants of the letter "A", which will be denoted by A_1, A_2, etc.

Let R be the original network, upon which the various versions of the patterns A_1, A_2, etc. are mapped, and let C be the coding field. The new network, R', is the union of of R and C, and the patterns in it will be denoted by an apex. Let us also assume, from here onwards, that the activation of the first node of C codes the information "it is an A".

In the learning stage, when presenting the different versions of A, the corresponding bit in the coding field will be switched on: A'_1 is thus a pattern which is identical to A_1 in R, while the only element active in C is the first one. Analogously, A'_2 is the same as A_2 in R, but it also has the first bit in the coding field on, as for all the possible versions of the letter "A". Due to the reinforcement of connections induced by Hebbian learning, even some mixed states (e.g., A_1 AND A_2) may have the first bit in the coding field on: the higher level system can thus easily classify these states as representing the letter "A".

A further advantage is that this system can discriminate between the mixed states with defined semantics and those with dubious semantics: in the case of a mixed state between A and B, for example, it is possible that the excitation-inhibition in play will turn on both bits in the coding field, or neither, generating a situation of uncertainty which can be easily detected.

The use of coding fields presents important differences with respect to the completely distributed approach. In fact, in this case, in addition to the distributed representation of the patterns we have introduced a local representation, where the activation of a single element (a "grandmother neuron", in neural jargon) represents a high level concept.

This method presents obvious advantages: the concentration of the information makes the results obtained more readable and better usable both for an external human observer and for other subsystems which make up the overall cognitive system.

This technique, however, presents also disadvantages, typical of local representations. One of these is the lack of fault tolerance: every fault in the cells of the coding field would cause the complete loss of the corresponding information. Moreover, one has to prescribe all the possible classes to be recognized, and to allow them room in the coding field.

In order to keep the advantages of abstract coding without loosing fault tolerance and spontaneous emergence of categories, the various levels of abstraction can be realized, in their turn, using distributed representations. For example, the high level coding of "A" and "a" can be the same, without corresponding to the switching on of a single bit, but to the activation of a distributed representation over a subnetwork at a higher conceptual level.

Moreover, the connections from the original network to the higher level field could also be one-directional: this could be achieved, for example, in a two-layered feedforward network.

The examples presented so far in this chapter dealt with networks without hidden units, where the meaning of the single nodes is therefore predefined. We have however seen, in the previous chapters, how important hidden units can be. In this case, their meaning is not given a priori, but comes out, often in a largely unexpected way, during the learning phase.

As we have seen, it is sometimes possible to attribute an explicit meaning also to the hidden nodes, *a posteriori*.

Some examples can be found in Chap. 5: for instance, in the case of the exclusive OR, the internal units compute logical functions (predicates) of the two input variables. In the case of the detector of "meaningful discontinuities", one hidden node computes the difference between the average signal levels to the right and to the left of the central point, while the other detects a discontinuity at the centre of the input window. It has also been proposed to automatically extract production rules from the analysis of neural networks applied to problems of "symbolic knowledge", like medical diagnosis (Fogelman, 1988: see the discussion below in Sect. 7.4): clearly, this hypothesis is based upon the possibility of attributing an explicit meaning to the network nodes.

However, in more complicated networks it is very difficult or impossible to attribute an explicit meaning to the action of the hidden units, whose way of functioning constitutes an essentially unreadable encoding (a "dynamical encription") of the characteristics of the proposed task.

To conclude this section, we would like to briefly comment upon two expressions which are sometimes used to describe aspects of neural network research: "knowledge without representation" (Varela, 1986) and "subsymbolic computation" (McClelland and Rumelhart, 1986).

The first expression refers to the fact that the modelling of cognitive processes using artificial systems capable of self-organization and self-creation of meaning (see Chap. 1) may take place in a manner largely independent of the system designer, and sometimes incomprehensible to the latter. This fact must be compared with the necessity of a detailed and precise representation of knowledge in classical AI systems.

However, the term "knowledge representation" is often used to indicate the main bone of contention between cognitivists and behaviourists. Behaviourists reject the scientific validity of referring to hypothetical internal representations, affirming that, in the formulation of psychological theories, only directly observable experimental quantities should be used. Cognitivists, on their part, stress the need to use representational states in the explanation of psychological behaviour, posing the problem of their adequate characterization.

From this point of view, both classical AI and connectionism lie on the "representationalist" side (Rumelhart and McClelland, 1986; Fodor and Pylyshin, 1988). Therefore, in order to avoid confusion, we prefer not to use the term "knowledge without representation" when discussing connectionist models.

The term "subsymbolic computation" is frequently used to describe the connectionist approach. It refers to the fact that, in distributed representations, a node is not associated with one particular symbol, being able to take part in the distributed representation of the various concepts. "Symbols" therefore emerge as activation patterns of lower level entities.

Whatever the intention of the inventors of this term, it might suggest a sort of hierarchical architecture, with a sub-symbolic layer dealing with low level information, and a symbolic layer devoted to high level cognitive tasks. However, as we have seen, neural networks are capable of handling also cases which involve a kind of knowledge which is often described as "symbolic", like in medical diagnosis or in natural language understanding systems.

In order to stress the fact that the dynamical systems approach is not limited to low level information processing tasks (like e.g. image and speech processing, or sensory-motor coordination) we prefer to avoid referring to the term subsymbolic computation, whereas the concept of distributed representation is widely used throughout this book.

7.3   The Role of Dynamics

The central theme of this book is that complex dynamical systems are useful and important for the comprehension and the artificial simulation of cognitive behaviour. It has also been suggested, in Chap. 2, that one of the main reasons for the re-awakening of interest in this approach is to be found in recent developments in the field of nonlinear dynamics.

Nevertheless, most of the models examined so far can also be described in other terms, where the dynamics plays a somewhat secondary role. Take, for example, the symmetrical Hopfield model, where the dynamics always leads to fixed points. This model is therefore an associative memory which, as the name suggests, associates a given input state with the corresponding output state. This task, however, can also be carried out using other techniques, which make no reference to dynamics but rather to algebraic and projection operator techniques (see Kohonen, 1984 and Bottini, 1980).

Also in the case of backpropagation networks the task carried out consists in associating an input state to an output state. Moreover, as we have already seen, in these networks the absence of feedback prevents the establishment of complex dynamics in state space.

One might therefore ask whether the role of dynamics is really essential. As mentioned in the introduction, the human brain is a nonlinear dynamical system, so that the interest of the neuroscientists for modelling dynamical neural networks is justified. However, researchers more oriented towards the artificial have to face the problem of comparing this latter approach with other associative techniques, which are not based upon dynamical systems.

As long as the decision regions are fairly regular, and the asymptotic states are represented by static attractors, geometrical methods can be used in place of dynamical ones. Geometrical and dynamical methods can be compared on the basis of their performance; however, dynamics is not the only possible solution, from a conceptual point of view.

If emphasis is placed upon the emergence-oriented approach instead of upon the control-oriented one, the adoption of a dynamical systems framework is fundamental, since it allows one to highlight the relationships between cognitive systems and other nonlinear systems, both natural and artificial, which are capable of self-organization.

Dynamics, however, can be fundamental also from the control-oriented viewpoint. We believe, in fact, that the substantial equivalence between dynamical and geometrical methods is limited to the case of systems having static attractors (i.e., fixed points) and relatively regular basins of attraction. If, however, more complex systems are considered, then the two approaches diverge.

Take, for example, fractal basins of attraction (see Chap. 3): in this case it is no longer possible to adequately approximate the behaviour of a dynamical system with regular decision regions.

Also the introduction of dynamical attractors, instead of fixed points, differentiates the two approaches. It is also worth pointing out that, in nonlinear dynamical models, truly dynamical attractors are the rule, not the exception.

Within the context of using dynamical attractors, one particularly stimulating problem is the role played by deterministic chaos in neural networks. There are, in fact, experimental results (Babloyantz and Destexhe, 1986; Skarda and Freeman, 1987) according to which, in electroencephalographic signals, low dimensional chaotic attractors have been discovered, which indicates a situation of deterministic chaos, as discussed in Chap. 3.

It would thus seem that, at least under certain circumstances, the human brain operates like a chaotic dynamical system. On the other hand, the presence of deterministic chaos in high-dimensional nonlinear systems is certainly not surprising (Sompolinsky, 1988; Basti and Perrone, 1989). The problem is that of understanding the role played by these chaotic states in cerebral information processing and the one which they could play in artificial cognitive systems.

An interesting proposal, presented by Skarda and Freeman (1987), is that deterministic chaos is an efficient brain reset mechanism. By studying the rabbit olfactory bulb both in laboratory and by means of a simulation model, they suggested that the recognition of various odours is to be associated with the relaxation of a neural network towards cyclic asymptotic states.

The chaotic state appears during exhalation, eliminating the memory of the preceding odour by activating, in a largely random manner, various elements of the network. The ground state of the network would therefore be chaotic, and only the presence of a particular odour as input, during inhalation, would drive the system to reach the limit cycle corresponding to that odour.

The establishment of a chaotic regime would provide the system with a certain "impartiality", and with an almost unbiased probability of reaching the various meaningful limit cycles.

The great advantage of deterministic chaos as a reset mechanism, with respect to brownian noise, derives from the fact that it is possible to switch it on or off almost instantaneously, through (a sequence of) bifurcations.

However, the question as to whether chaotic states may be more deeply involved in information coding and processing remains open (Nicolis, 1986; Oi, 1987; Basti and Perrone, 1989).

In this context, it is also interesting to underline the fact that chaotic systems have the capability to generate information. Consider, for instance, a system with a sensitive dependence on its initial conditions. It is well known that every observation we can make has a finite precision. Let us now consider two different initial states, the distance between them being less than our resolving power: they would therefore seem to coincide. At time $t = 0$ we thus see them as equal.

During the course of dynamical evolution, in a system of this type, the two trajectories separate such that, at a certain time, the distance between the two states becomes greater than the resolving power of our observations: we can now see two states instead of a single one. The system has generated information over the course of time, allowing the discrimination between states which initially appeared to be coincident.

This capacity to "generate information" may be an important characteristic of cognitive systems.

In any case, the definition of the role played by chaotic states in neural dynamics is still at an early stage and will be the object of numerous studies and hypotheses in the near future.

7.4 On the Limits of the Dynamical Approach

So far, we have stressed the advantages of the dynamical approach, which avoids some of the main problems in classical AI, as shown by the study of the various specific models carried out in Chaps. 4 to 6. We must now outline a critical examination of the main problems raised by this type of approach.

First of all, it must be stressed that the dynamical approach is still in an embryonic stage, in which it has been possible to demonstrate that there are dynamical systems with interesting properties, but the problems faced so far have almost always been small, strongly stylized problems. The application of these dynamical models to real size problems is just in a beginning phase, and it will require a great deal of theoretical developments.

One fairly obvious difficulty of dynamical systems regards the handling of long inference chains. Classical AI systems are built precisely for dealing with this type of reasoning, and it is not realistic to expect that, in this sector, dynamical systems will be as efficient and powerful.

A limit of classical AI methods, however, is that they tend to express every cognitive operation as an inference chain, whereas there are some which may be better described in other terms.

Connectionist systems do, in fact, appear to be more suitable than "classical" ones for dealing with various forms of knowledge which, from an anthropomorphous point of view, are considered "low level" ones, like for example analysis and recognition of signals, images and speech, speech synthesis, sensor-movement coordination in robots, etc.

In these sectors, the introduction of inferential models becomes problematic due to the presence of two different types of difficulty. On one hand, it seems to be unnatural to human experts, and this creates severe knowledge engineering problems. On the other hand, the signals handled are affected by noise and, as we have seen, inferential systems are fragile, whereas dynamical systems are more robust and show a greater capacity for generalization.

It is not by chance, therefore, that most connectionist system applications have so far been directed towards the treatment of low level knowledge. In these sectors there are already interesting demonstrations of the capabilities of neural networks: for example, in image analysis for handwritten character recognition (Guyon et al., 1986) and quality control (a working system was presented at the IEEE ICNN exhibition in 1988, see Glover (1988)), in sonar signal recognition (Gorman and Sejnowski, 1988) and in speech synthesis (Sejnowski and Rosenberg, 1987) and recognition (Kohonen, 1988b; Bengio et al., 1989).

Moreover, neural networks have proven effective in several tasks relating to statistical data analysis (Gallinari et al., 1988).

It has not been possible, so far, to prove that neural networks are superior with respect to the more usual statistical methods in the fields of signal recognition and analysis (Lippman, 1987). It is, however, interesting from a conceptual point of view to discover that self-organized systems, which learn by examples, are capable of obtaining results which are comparable with those obtained using the best statistical methods (like hidden Markov models: Bengio et al., 1989). Also, from an applications oriented point of view, it is important to note that the time needed to train a neural network capable of carrying out a certain recognition can be much less than that required to adjust statistical algorithms for the same task.

It can therefore be said that, today, neural networks already represent an important instrument for low level knowledge processing. It is interesting, however, to examine the problem of higher level knowledge: what is the significance of dynamical systems for the execution of cognitive tasks which require, to be carried out by a human being, a mixture of high level theoretical knowledge and specific experience?

The state of the art in the application of the dynamical approach to high level knowledge is at a much lower stage of development than that for low level knowledge. There are, however, some experimental systems which allow early considerations to be made (Bounds et al., 1988; Collins et al., 1988; Dutta et al., 1988; Fogelman et al., 1988; Surkan, 1988).

In all these systems the case under examination is described using a certain set of variables – like e.g. symptoms in the case of medical diagnosis, information about the financial situation of the applicant in the case of bank loan concession, etc. This set of variables constitutes the input pattern, while the output is given by the corresponding diagnosis, or decision. The network learns a set of symptom-diagnosis associations; its ability to generalize then allows it to operate also on cases which are not included in the training set.

In contrast with expert systems and knowledge engineering, here there is no attempt to obtain from the expert an explicit formulation of his own problem solving method, but the results are obtained by examining a set of examples. To produce a system of this kind, therefore, it is not necessary to perform knowledge engineering together with an acknowledged expert, but rather to have a large number of cases. With respect to other systems which learn by examples, based upon the inferential method of AI, an important advantage of connectionist systems lies in their robustness with respect to possible inconsistencies present in the training set.

Neural diagnostic systems seem to operate in a more "holistic" manner than inferential ones: in fact, from their "creator's" point of view, they learn to associate entire input patterns with the corresponding output decisions. The "microscopic" operations of the extraction of particular subpatterns and the "logical" combinations of features spontaneously develop within the system, and do not require explicit coding.

A serious drawback of the use of dynamical systems for handling high level knowledge lies in their limited explanation capabilities. As we have seen, although it is sometimes possible to understand the contribution of the operation of a particular unit, this is not usually the case. The limitations in explanation capabilities do not affect the performance level, but they can make it difficult to accept to rely upon the network's conclusions. The possibility of explaining the reason why an automatic system has reached a certain conclusion is one strong point of expert systems: actually, such topic is such important that Kodratoff (1989) recently proposed to identify AI with the "science of explanations".

We also note that most diagnostic systems developed so far, based upon neural networks, use multilayered feedforward networks. As we know, these systems undergo a kind of learning which is essentially statistical; its result reflects the frequency with which the associations between input and output patterns in the training set appear. The statistical nature of learning in dynamical networks constitutes one of the limits of this type of approach. In particular, cases where only a few examples are available, upon which a profound level of reasoning must be developed, are not suited for it.

A further point is worth stressing. Classical AI systems rely on the use of structured expressions: a given set of "atoms" is defined, along with composition rules to generate more complex expressions. No such automatic way of building complex structured expressions exists so far in connectionist models (Fodor and Pylyshin, 1988). In this case, the structure of the network itself must be tailored for the purpose being considered.

Another serious problem for the development of dynamical systems for expert tasks is the introduction of previous domain knowledge.

The most fascinating hypothesis, the most tempting dream is the one of a completely general system, capable of learning anything at all by examples. The properties of discrimination, generalization and robustness of such a system should provide it with the capabilities, given an arbitrary series of examples, to develop a sensible "theory" about the domain. The word "theory" is intended here as a representation which allows a reasonably good performance even in cases other than those in the training set; naturally, the value of the theory will strongly depend upon the quality of the training set, which constitutes the only source of knowledge for this type of system.

The beauty of this hypothesis lies in the fact that it is based exclusively upon self-organization properties induced by examples. An artificial system of this kind would really be capable of general intelligence. However, it does not exist.

Actually, there is a natural system which has developed through increasingly complex forms of self-organization, and which is capable of general intelligence, that is, the human brain. However, we must take into account the different timescales involved in cerebral self-organization processes.

Learning, in the human case, is a process which lasts for some decades, while biological evolution has been operating for millions of years. The brain with which we are born possesses a structure which represents the memory of this very long history. It is well-known that the brain is not initially a homogeneous network which trains itself only from examples during our lifetime. The brain shows some evident structures and major information flows can be be identified (Changeux, 1983).

This structuring is not complete: there is not, in fact, enough DNA to code all the synapses. What seems to occur in the brain, during the lifetime of an individual, are circumscribed self-organization phenomena, within the framework of a genetically pre-defined general structure. One therefore has an interplay between self-organization and an a priori defined structure.

It is true that also this structure is, in its turn, the product of self-organization processes, but these are so slow that they do not appear over periods of time of the order of a human lifetime. If we are building artificial systems capable of cognitive performance, we ourselves – the designers – must play the role which is carried out by biological evolution, introducing suitable structuring elements.

Moreover, in the case of human societies, "superior" performance does not only reflect the biological structure of the nervous system, but also requires cultural transmission mechanisms. Cultural evolution is much faster than biological evolution, and is therefore capable of much more rapid adaptation. Cultural transmission takes place only partially through the mere presentation of examples: it necessarily involves other learning mechanisms, as the explicit transmission of concepts, methods and techniques.

Consider, for example, typical expert tasks, such as medical diagnosis, chemical analysis, tracing of faults in complex apparatus, interpretation of geophysical data, etc. These require – besides field training which can be modelled as learning by examples – also a significant quantity of explicit knowledge, transmitted through books, lessons, manuals, and diagrams.

In order to obtain artificial neural systems which are effective in facing complex problems of this type it is thus necessary to be able to provide them, in one form or another, with explicit knowledge, specific to the field of application of interest. So far, there are only partial and limited solutions to the introduction of this type of knowledge in neural networks.

A very important technique for introducing previous domain knowledge is naturally the adequate choice of input variables. This operation is necessary in the design of any artificial system, because access to relevant information is obviously a condition for any significant performance. It is also an operation which is external to the system itself, since it must be performed by the system designer.

Anyway, while any artificial system requires a pre-treatment or at least a preliminary choice of the information with which it is supplied, a strategy based upon very sophisticated pre-processing merely shifts the difficulties to another module of the system: how to carry out a satisfactory pre-processing still has to be understood.

Another possible route, also suggested by biology, is that of pre-wiring a portion of the connections. Let us consider, for example, a three-layer network: in this case, not all of the input elements will be connected to all those in the intermediate layer, some of these connections being eliminated right from the start. In a sense, this corresponds to the generation of a certain number of "grandmother neurons", encharged with calculating largely predefined combinations of the inputs. This set of techniques may be called "connections engineering", by analogy with knowledge engineering.

Another possibility is based upon the a priori translation of the knowledge into the architecture of a modular system. In fact, complex cognitive tasks generally require the development of complex, hierarchical and possibly heterogeneous cognitive systems: domain knowledge can then be reflected in the design of the system modules and their interactions.

Let us first of all examine the problems associated with connection engineering. Although we have so far maintained its necessity for theoretical reasons, it is interesting to note that several empirical attempts have been directed precisely towards this aim. A case documented in the literature is that of a system for the diagnosis of abdominal pain (Fogelman et al., 1988) where initial trials were held using an "uninformed" back propagation network. The results, however, were not completely satisfactory, so that a different architecture was tried, with an extra intermediate layer, in which various "syndromes" (e.g., urinary diseases, gynaecological diseases) were represented.

A priori knowledge was introduced by reducing the number of connections, in order to leave only those between units which represent concepts known to be interrelated. For example, the connection between "gastric or duodenal pain" and the diagnosis "ulcer" was left, because it is known that an ulcer can cause abdominal pain and it is often situated in the stomach or the duodenum. In this way the performance of the system considerably improved, reaching about 70% in the generalization stage.

The utility of the reduction in the number of connections can be understood by recalling that the learning procedure is based upon minimizing a cost function in weight space. In this way, therefore, the number of degrees of freedom of the space where optimization takes place is reduced, thus reducing also the risk of becoming trapped in local minima. In fact, in complex systems such as nonlinear neural networks, the number of local minima rapidly increases with the number of degrees of freedom.

There are various cases of signal and low-level knowledge processing where it is possible to reduce the number of degrees of freedom by building the connections with a problem dependent topology. For example, in image analysis it can be useful to construct an intermediate layer, placed immediately behind the "retina", in which every element is connected to only a small portion of the retina (Le Cun, 1989). This choice is based upon a priori knowledge deriving from the image analysis field: it is, in fact, well-known that many of the initial steps in image processing involve purely local operations on points belonging to only a small neighbourhood.

In some cases it is possible to reduce the number of degrees of freedom without severing the connections, but by imposing that they are not all independent ones (Le Cun, 1988). For example, the symmetry properties of the problem may suggest making some of the connections equal, which are then forced to evolve in the same manner.

The techniques for introducing a priori knowledge into neural networks currently constitute one of the most interesting lines of theoretical research. The methods mentioned so far, together with those presented in the literature, nevertheless seem once again to be more suitable for low level rather than for high level knowledge. In the latter case it is, in fact, quite difficult to have symmetry properties in the connections, or "local" interactions uniformly spread over a whole subset of the network (as in the example of local processing of images on the retina).

It would therefore be necessary, each time, to carry out a detailed and precise connection engineering task, which would be rather unnatural and perhaps even more difficult than "knowledge engineering" in rule form.

One might ask whether the emphasis placed upon a priori knowledge reintroduces all the problems of knowledge engineering, discussed in Chap. 2. The answer is negative, because the really difficult and unnatural part of that process consists in the complete formalization of the domain knowledge, whereas the introduction of partial and limited knowledge ought to be simpler. One could thus imagine a stage of rapid knowledge engineering followed by an improvement in performance obtained through learning by examples.

The difficulties of connection engineering point to the need for integrating the dynamical systems approach and the classical approach; such a need had already been stressed previously on the basis of the complementary characteristics of the two approaches. Although there are still very few studies focusing upon the problems of such integration, the main proposals may be grouped together according to the degree of interaction between the two approaches.

A relatively cautious proposal is that of "heterogeneous systems", i.e. systems composed of two or more modules, some symbolic and others dynamical, which specialize in different tasks. This might be the case, for example, of a hypothetical speech comprehension system where a neural network could perform speech recognition, by formulating "recognition hypotheses" to be sent to a symbolic module for syntactic and semantic analyses.

Such a "loose coupling" solution is the least dangerous one, since the two approaches are kept separate, interacting only through the exchange of data: the output of a given module is (a part of) the input of another module.

Different couplings between the two approaches are also possible: one attempt is the proposal (Hendler, 1989) of exchanging activation between a connectionist network and a semantic network of the "marker passing" type.

Another different suggestion is that of using the results which can be achieved by training neural networks in order to provide hints and suggestions for the knowledge engineering phase, which is, however, not automatic. Some possibilities which have been suggested along this line (Fogelman et al., 1988) include the interpretation in terms of high level descriptors of the structures which spontaneously develop in neural nets when trained in specific tasks, and the deriving of some rules for expert systems from an analysis of the connection scheme after supervised learning.

A different approach is used in systems which might be called "hybrid", because they integrate inferential and dynamical techniques in the same module. This line of research is particularily interesting for at least two fundamental reasons:

- it is based on a very tight coupling between the inferential and the dynamical paradigms, thus allowing a thorough investigation of the interplay between the two;
- it allows a straightforward introduction of previous knowledge, with conventional knowledge engineering techniques.

The essential idea behind this line of research is that of superimposing a "selectionist" (Steels, 1987) dynamics onto a rule based system. The best known example in this field is that of classifier systems, whose properties were studied in Chap. 6.

In these systems, in fact, the introduction of domain knowledge (the rapid knowledge engineering stage) is straightforward, and consists of the direct introduction, into the initial classifier population, of some a priori rules. The learning by examples stage then serves the purpose of improving the knowledge base of the system.

Moreover, classifier systems present a very strong interaction between inferential and dynamical aspects (consider, for instance, the solutions discussed in Sect. 6.4), thus allowing an analysis of the advantages and of the problems posed by this tight coupling of the two approaches.

A very powerful feature of classifier systems is their capability of modifying their own structure in order to deal with the task at hand.

Alongside these features, which render them extremely interesting, classifier systems show, in their present form, severe limitations.

First of all, it is necessary to obtain a better understanding of their learning mechanisms. It often happens, for example, that their performance critically depends upon several parameters, and it is necessary to find the most important ones, as a function of the characteristics of the task to be carried out. Analyses of the dynamics and the stability of learning, partially summarized in the previous chapter, have precisely this orientation.

Since these systems involve a large number of parameters, it is extremely unlikely that there is a set of values and choices suitable for a wide range of different cases. In order to develop a powerful and flexible cognitive system, based upon classifiers, one or more B-brains (Minsky, 1986) need to be introduced. A B-brain is, in its turn, a cognitive system, which monitors the performance of another cognitive system (A-brain), directly connected to the external environment:

EXTERNAL ENVIRONMENT ↔ A-BRAIN ↔ B-BRAIN

A B-brain associated with a classifier system could, for example, check its performance and decide upon whether to change some of the choices. It could also develop a more detailed analysis of the causes of faults: recalling, for instance, the example discussed in Sect. 6.4, it could realize that the rule population has sub-divided into subpopulations, and that the forces are equalizing, and thus introduce suitable correctives in order to avoid problems which, as we know, can manifest themselves in these cases.

In fact, our research upon the dynamical behaviour and the stability of solutions of classifier systems are directed towards a definition of the functionalities of suitable B-brains.

We have already mentioned that the applications of classifier systems still concern highly simplified microworlds, and not real size problems. One of the main reasons for that is certainly to be sought in the poor representation scheme based upon binary messages of fixed length. Although this is not a limitation in principle, it is a serious limitation in practice. It is well known (Lenat, 1984) that representations of suitable level are required in order to make mutation mechanisms effective.

The power of representation of classifier systems could be improved along several directions, including:

- enlarging the range of allowable values for each cell (e.g., real instead of boolean variables);
- introducing operators (more powerful than "don't care") which compute the action from the condition (instead of simply associating them);
- abandoning the string representation and building the reward and genetic mechanisms upon more sophisticated languages (e.g., OPS5 or Prolog).

Clearly, the use of more elaborate classifier systems also requires the adoption of other genetic operators, and the abandoning of a strict analogy with biological genetics.

Note how the complicating of the individual classifiers shifts the "Michigan approach" to classifier systems somewhat closer to the "Pittsburgh approach". Actually, the main difference among the two is to be found in the granularity of the individuals upon which genetics operates: simple rules in the former case, sets of rules in the latter case.

The utilization of more sophisticated and expressive classifiers than the present ones may lead to an interesting evolution of classifier systems, in a sense which deserves some attention. Classifier systems are based upon a sort of population dynamics, where the individuals (the rules) are fairly simple. If we make their structure increasingly powerful and expressive, we induce population dynamics among rather complex agents, each of which is specialized in carrying out certain operations. In this way, a kind of "society of mind" (Minsky, 1986) emerges, composed of various specialized agents which interact with one another.

Classifier systems seem to provide a good starting point, a "laboratory" for experimenting with this "society of mind" model: an interesting aspect is that the various agents communicate explicitly through messages in the message list, and that there is a dynamics which is capable of measuring the "value" of every agent with respect to the system. There is only one legal tender for the whole society, which is "strength". Another important aspect of this proposal is obviously the introduction of mutation mechanisms and of the possibility of developing new agents, starting from the existing community.

Another suggestion, which in some respects moves in the same direction as that mentioned above, is that of integrating neural networks and genetic algorithms into a single system (Muehlenbein, 1988).

Amongst the various dynamical models proposed for AI, other than neural networks, one should recall the idea, proposed by Luc Steels and already analyzed in Chap. 3, of using simplified dynamical systems (typically, cellular automata) for representing common sense. This method is fairly straightforward when considering motion and spatial skills, as in the example of a robot moving in a changing environment by following the concentration gradient of an "odour" emitted by the target.

It is also possible, however, to apply this technique to the study of apparently more abstract problems, by placing emphasis upon geometrical or motion aspects of the problem or of the solution procedure. For example, the eight-puzzle problem, usually presented as an abstract tree-searching problem, has been approached and solved by placing the accent upon the dynamics of the active forces (Steels, 1988): an attractive force, which recalls the out-of-place pieces towards their "natural" location, repulsive forces which prevent overlap between the pieces and attractive forces which form chains of pieces with the correct relative positions.

This approach is thus based upon the development of models which simulate, in a very simplified manner, physical processes, and also upon attractive forces emanating from the goal state, in some kind of suitable representation. From the point of view of simulating human cognitive properties, this link between abstract tasks and concrete operations and relationships is extremely interesting.

From the point of view of an artificial system, the use of a simplified simulator is motivated by the fact that more realistic simulators (for example, fluid dynamics models based upon Navier-Stokes equations) are often impracticable, requiring too many data and too much computing time, and often fail to highlight the important aspects of a solution.

Note how this hypothesis, although based upon a simplified physics, differs from the "naive physics" approach (De Kleer and Brown, 1984) which attempts to describe qualitative physical properties using symbolic rules, without any use of dynamical systems.

So far, in this volume we have stressed the "fragility" of the logic approach compared to the robustness of the dynamical approach. This latter property is shown by the tolerance of networks to noise, errors and damage. It should be noted, however, that also dynamical systems present quite serious problems of fragility, although of a different nature, since their performances can be extremely sensitive to the values of some parameters; this, in a system with a lot of modifiable parameters, may lead to bad results.

If one is interested in reaching significant levels of performance for complex tasks, it is often necessary to "strengthen" the system by dissecting it into modules, and suitably regulating the interactions between them. The modularization, which also has the advantage of simplifying interactions between inferential modules and dynamical ones in a possible heterogeneous system, has the characteristic of limiting the self-organization of the network, prescribing its structure externally. This structuring, however, is often necessary for reaching high performance levels: the designer at his drawing board has to play the role which, in the case of the human brain, was played by biological evolution.

References

The present reference list is not meant to be complete. The literature about dynamical systems for AI is growing extremely fast, and any such attempt would be condamned to failure.

It is also beyond our purpose to give a complete reference list up to, say, the end of 1988. Such an attempt would have compelled us to prepare a list of some thousand papers, most of them being probably of limited interest for the reader.

We have thus confined ourselves to provide a reasonably wide list, which we consider a good starting point for one's own explorations in this fascinating field. The emphasis placed about some authors reflects the choices of the volume, and in particular the central role of dynamical systems. However, references to some important works which are not explicitly discussed in the text, but which regard closely related topics, are also included.

When there exist several papers describing essentially the same results, we have usually provided the quotation of only one of them, typically the most detailed one (e.g., a book).

Ackley, D.H., Hinton, G.E., Sejnowski, T.J. (1985): Cognitive Sci. 9, 147-169

Amari, S. (1972): IEEE Trans. C-21, 1197-1206

Amit, D.J., Gutfreund, H., Sompolinsky, H. (1985a): Phys. Rev. A 32, 1007-1018

Amit, D.J., Gutfreund, H., Sompolinsky, H. (1985b): Phys. Rev. Lett. 55, 1530-1533

Amit, D.J., Gutfreund, H., Sompolinsky, H. (1986): Information storage in neural networks with low levels of activity. Preprint

Anderson, D.Z., Erie, M.C. (1987): Opt. Engng. 26, 434-444

Anderson, J.A. (1972): Math. Biosci. 14, 197-220

Anderson, J.A. (1986): Cognitive capabilities of a parallel system. In: Bienenstock, E. et al. (1986)

Anderson, J.A., Silverstein, J.W., Ritz, S.A. (1977): Psychol. Rev. 84, 413-451

Anderson, J.A., Rosenfeld, E. (eds.) (1988): Neurocomputing. Foundations of research. Cambridge, MA: MIT Press

Arecchi, F.T. (1986): Caos e ordine nella fisica. In: Serra, R., Zanarini, G.: Tra ordine e caos. Bologna: Clueb

Atlan, H. (1985): Functional self-organization and creation of meaning. In: Proceedings Cognitiva 85 (Forum, pp. 117-119). Paris: Cesta

Atlan, H. (1987): Physica Scripta 35, 123

Atlan, H., Fogelman-Soulié, F., Salomom, J., Weisbuch, G. (1981): Cybern. Systems 12, 103

196   References

Axelrod, R. (1987), in: Davis (1987), pp. 32-41
Babloyantz, A., Destexhe, A. (1986): Proc. Natl. Acad. Sci. USA *83*, 3513-3517
Baiardi, F., Vanneschi, M. (1987), in: Treleaven and Vanneschi (eds.) (1987), pp. 1-34
Ballard, D.H., Hinton, G.E., Sejnowski, T.J. (1983): Nature *306*, 21-26
Barr, A., Feigenbaum, E.A. (1981): The handbook of artificial intelligence. Los Altos, CA: Morgan Kaufmann
Barto, A.G., Sutton, R.S, Anderson, C.W. (1983): IEEE Trans. Systems, Man and Cybern. *13*, 834-846
Basti, G., Perrone, A. (1989): The cognitive function of deterministic chaos in neural networks. Preprint
Belew, R.K., Forrest, S. (1988): Machine Learning *3*, 193-223
Bell, T. (1988): Sequential processes using attractor transitions. In: Touretzky et al. (1988)
Bergé, P., Pomeau, Y., Vidal, Ch. (1984): L'ordre dans le chaos. Paris: Hermann
Bengio, Y., Cardin, R., De Mori, R., Merlo, E. (1989): Comm. ACM *32*, 195-199
Bienenstock, E., Fogelman Soulié, F., Weisbuch, G. (eds.) (1986): Disordered systems and biological organization. Heidelberg: Springer
Bottini, S. (1980): Biol. Cybern. *36*, 221-228
Booker, L.B. (1988): Machine Learning *3*, 161-192
Bounds, D.G., Lloyd, P.J., Mathew, B., Waddell, G. (1988): A multilayer perceptron for the diagnosis of low back pain. In: Proceedings of the IEEE International Conference on Neural Networks, San Diego, CA, vol.II, pp. 481-490
Burks, A.W. (ed.) (1970): Essays on cellular automata. Urbana: University of Illinois Press
Caianiello, E.R. (1961): J. Theor. Biol. *1*, 209
Caianiello, E.R. (1965), in: Caianiello, E.R. (ed.): Cybernetics of neural processes. Rome: Consiglio Nazionale delle Ricerche, pp. 333-360
Caianiello, E.R. (1986), in: Palm, G., Braitenberg, V. (eds.): Brain theory. Heidelberg: Springer, pp. 147-160
Caianiello, E.R. (ed.) (1988): Parallel architectures and neural networks. Singapore: World Scientific
Carnevali, P., Patarnello, S. (1987): Europhys. Lett. *4*, 1199-1204
Carpenter, G.A., Grossberg, S. (1987): Computer vision, graphics and image processing *37*, 54-115
Ceruti, M. (1986): Il vincolo e la possibilitá. Milano: Feltrinelli
Chaitin, G. (1975): Sci. Amer. *232*, 47
Changeux, J.P. (1983): L'homme neuronal. Librairie Artème Fayard
Cohen, M.S. (1986): Appl. Optics *25*, 2288-2294
Collins, E., Ghosh, S., Scofield, C.L. (1988): An application of a multiple neural network to emulation of mortgage underwriting judgements. In: Proceedings of the IEEE International Conference on Neural Networks, San Diego, CA, vol.II, pp. 459-466
Compiani, M., Montanari, D., Serra, R., Valastro, G. (1988): Classifier systems and neural networks. In: Caianiello, E.R. (ed.) (1988), pp. 105-118
Compiani, M., Montanari, D., Serra, R., Simonini, P., Valastro, G. (1989a): Dynamical systems in artificial intelligence: the case of classifier systems. In: Pfeifer, R. et al. (eds.): Connectionism in perspective. Amsterdam: Elsevier, pp. 331-340
Compiani, M., Montanari, D., Serra, R., Simonini, P. (1989b): Asymptotic dynamics of classifier systems. In: Schaffer, D.J. (ed.): Proc. III Intl. Conf. on Genetic Algorithms. Los Altos: Morgan Kaufmann, pp. 298-303
Compiani, M., Montanari, D., Serra, R. (1989c): Learning and bucket brigade dynamics in classifier systems. Physica D (in press)
Cooper, L.N. (1973), in: Lunquist, B., Lunquist, S. (eds.): Proceedings of the Nobel Symposium on Collective Properties of Physical Systems. New York: Academic Press
Cortes, C., Krogh, A., Hertz, J.A. (1986): Hierarchical associative memories. Nordita preprint 86/19 S

Crutchfield, J.P., Farmer, J.D., Hubermann, B.A. (1982): Phys. Rep. *92*, 45

Davis, L. (1987): Genetic algorithms and simulated annealing. London: Pitman

Davis, L., Steenstrup, M. (1987), in: Davis, L. (1987), pp. 1-12

De Jong, K. (1988): Machine Learning *3*, 121-138

De Kleer, J., Brown, J.S. (1984): Artificial Intelligence *24*, 7-83

Denker, J.S. (1986): Neural networks for computing. New York: American Institute of Physics

Denker, J., Schwartz, D., Wittner, B., Solla, S., Howard, R., Jackel, L., Hopfield, J. (1987): Complex Systems *1*, 877-922

Dutta. S., Shekhar, S. (1988): Bond rating: a non conservative application of neural networks. In: Proceedings of the IEEE International Conference on Neural Networks, San Diego, CA, vol.II, pp. 443-450

Eckmiller, R., von der Malsburg, C. (1988): Neural computers. Heidelberg: Springer

Elman, J.L., Zipser, D. (1987): Learning the hidden structure of speech. ICS Report 8701. San Diego: Institute for Cognitive Science

Fahlman, S.E., Hinton, G.E. (1987): Computer, January 1987, 100-109

Farhat, N.H., Psaltis, D., Prata, A., Paek, E. (1985): Appl. Optics *24*, 1469-1475

Farmer, J.D. (1982): Dimension, fractal measure and chaotic dynamics. In: Haken, H. (ed.): Evolution of order and chaos. Heidelberg: Springer

Farmer, J.D., Packard, N.H., Perelson, A.S. (1986): Physica *22* D, 187-204

Fasano, F. (1988): Proprietá computazionali di reti dinamiche. Doctoral dissertation, Physics Dept., University of Bologna

Feldman, J.A., Ballard, D.H. (1982): Cognitive Science *6*, 205-254

Fischer, K. (1983): Phys. Stat. Sol. (b) *116*, 357-414

Fodor, J.A., Pylyshyn, Z.W. (1988): Connectionism and cognitive architecture: a critical analysis. Cognition *28*

Fogelman Soulié, F., Goles Chacc, E., Weisbuch, G. (1982): Bull. Math. Biol. *44*, 715

Fogelman Soulié, F. (1985): Brains and machines: new architectures for tomorrow? In: Proceedings Cognitiva 85 (Forum, pp. 122-128). Paris: Cesta

Fogelman Soulié, F. (1986), in: Bienenstock et al. (1986), pp. 85-100

Fogelman Soulié, F., Gallinari, P., Le Cun, Y., Thiria, S. (1988): Network learning. In: Kodratoff, Y., Michalski, R. (eds.): Machine learning, vol.III (in press)

Fukushima, K. (1975): Biol. Cybern. *20*, 121-136

Fukushima, K., Miyake, S. (1980): Biol. Cybern. *36*, 193-202

Fukushima, K. (1988): Neural Networks *1*, 119-130

Gabor, D.(1969): IBM J. Res. Dev. *13*, 156-159

Gallinari, P., Thiria, S., Fogelman-Soulié, F. (1988): Multilayer perceptrons and data analysis. In: Proceedings of the IEEE International Conference on Neural Networks, San Diego, CA, vol.I, pp. 391-400

Gardner, M. (1983): Wheels, life and other mathematical amusements. New York: Freeman

Glover, D.E. (1988): Neural nets in automated inspection. Synapse connection *2* (6), 1-13

Goldberg, D.E. (1985): Dyanmic system control using rule learning and genetic algorithms. Proc. IJCAI-*85*, 588-592

Goldberg, D.E., Holand, J.H. (1988): Machine Learning *3*, 95-99

Gorman, R.P., Sejnowski, T.J. (1988): Neural Networks *1*, 75-90

Graf, H.P., Jackel, L.D., Howard, R.E., Straughn, B., Denker, J.S., Hubbard, W., Tennant, D.M., Schwartz, D. (1986), in: Denker, J.S. (1986), pp. 182-187

Grefenstette, J.J. (ed.) (1985): Proc. I Intl. Conf. on Genetic Algorithms and their Applications. Pittsburgh: Lawrence Erlbaum

Grefenstette, J.J. (ed.) (1987): Proc. II Intl. Conf. on Genetic Algorithms and their Applications. Cambridge: Lawrence Erlbaum

Grossberg, S. (1980): Psych. Rev. *87*, 1-51.

Gutfreund, H. (1986): "Neural networks with hierarchically correlated patterns", NSF-ITP-86-151

Gutfreund, H. (1987): Statistical mechanics of neural networks. Lecture at the French Annual Meeting of Statistical Mechanics

Guyon, I., Personnaz, L., Siarry, P., Dreyfus, G. (1987): Engineering applications of spin glass concepts. Preprint

Haken, H. (1978): Synergetics. Heidelberg: Springer

Haken, H. (ed.) (1980): Dynamics of synergetics systems. Heidelberg: Springer

Hebb, D.O. (1949): The organization of behaviour. New York: Wiley

Hecht-Nielsen, R. (1988): Neural Networks 1, 131-140

Hendler, J. (1989): Problem solving and reasoning: a connectionist perspective. In: Pfeifer, R. et al. (1989)

Hillis, W.D. (1985): The connection machine. Cambridge: MIT Press

Hinton, G.E., Anderson, J.A. (eds.) (1981): Parallel models of associative memory. Hillsdale, NJ: Lawrence Erlbaum

Hinton, G.E., Sejnowski, T.J. (1986): Learning and relearning in Boltzmann machines. In: McClelland et al. (1986)

Hogg, T., Huberman, B. (1984): Proc. Natl. Acad. Sci. USA, 81, 6871-6875

Hogg, T., Huberman, B. (1985): Phys. Rev. A 32, 2338-2346

Holland, J.H. (1975): Adaptation in natural and artificial systems. Ann Arbor: University of Michigan Press

Holland, J.H. (1986): Escaping brittleness. In: R.S. Michalski et al. (eds.) (1986): Machine learning. An artificial intelligence approach, vol.II. Los Altos, CA: Morgan Kaufmann, pp. 592-623

Holland, J.H., Holyoak, K.J., Nisbett, R.E., Thagard, P.R. (1986): Induction. Cambridge, MA: MIT Press

Hopfield, J.J. (1982): Proc. Natl. Acad. Sci. USA, 79, 2554-2558

Hopfield, J.J. (1984): Proc. Natl. Acad. Sci. USA, 81, 3088-3092

Hopfield, J.J., Tank, D.W. (1985): Biol. Cybern. 52, 141-152

Hopfield, J.J., Tank, D.W. (1986), in: Bienenstock et al. (1986), pp. 155-170

Huberman, B.A., Hogg, T. (1984): Phys. Rev. Lett. 52, 1048-1051

Huberman, B.A.: Collective computation and self-repair. In: Proceedings Cognitiva 85 (pp. 395-400). Paris: Cesta

Jen, E. (1986): J. Stat. Phys. 43, 219-265

Kauffmann, S.A. (1970): Behaviour of randomly constructed genetic nets: binary elements nets. In: C.H. Waddington (ed.): Towards a theoretical biology, vol.3. Edinburgh: Edinburgh University Press, pp. 18-37

Kauffmann, S.A. (1984): Physica 10 D, 145

Kauffmann, S.A. (1985): Selective adaption and its limits in automata and evolution. In: Demongeot, J., Goles, E., Tchuente, M. (eds): Dynamical systems and cellular automata. London: Academic Press

Keirstead, W.P., Huberman, B.A. (1986): Phys. Rev. Lett. 56, 1094-1097

Kinzel, W. (1985), in: Haken, H. (ed.): Complex systems – operational approaches. Heidelberg: Springer

Kinzel, W. (1985): Z. Physik B 60, 205-213

Kodratoff, Y. (1989): Enlarging symbols to more than numbers. In: Pfeifer, R. (1989)

Kohonen, T. (1972): IEEE Trans. Computers 21, 353-359

Kohonen, T. (1984): Self-organization and associative memory. Heidelberg: Springer

Kohonen, T. (1988a): Neural Networks 1, 3-16

Kohonen, T. (1988b): The neural phonetic typewriter. Computer, march 1988

Lashley, K.S.: In search of the engram. In: Society of experimental biology symposium, No.4. Cambridge: Cambridge University Press. Reprinted in: Anderson, J.A., Rosenfeld, E. (1988)

Le Cun, Y. (1988), in: Touretzky, D. et al. (1988)

Le Cun, Y. (1989): Generalization and network design strategies. In: Pfeifer, R. (1989)

Lee, Y.C., Doolen, G., Chen, H.H., Sun, G.Z., Maxwell, T., Lee, H.Y., Giles, C.L. (1986): Physica 22 D, 276-306

Lenat, D.B. (1984): The role of heuristics in learning by discovery: three case studies. In: Michalski, R.S. et al. (1984)

Lippman, R.P. (1987): IEEE ASSP Magazine, pp. 4-22

Little, W.A. (1974): Math. Biosci. 19, 101-120

Little, W.A., Shaw, G.L. (1978): Math. Biosci. 39, 281-290

Malferrari, L., Serra, R., Valastro, G. (1989): Backpropagation in noisy signal processing. Enidata internal report

von der Malsburg, C., Bienenstock, E. (1986), in: Bienenstock, E., Fogelman Soulié, F., Weisbuch, G. (eds.): Disordered systems and biological organization. Heidelberg: Springer, pp. 241-246

McClelland, J.L., Rumelhart, D.E., the PDP Research Group (1986): Parallel distributed processing. Cambridge, MA: MIT Press

McCorduck, P. (1979): Machines who think. New York: Freeman

McCulloch, W.S., Pitts, W. (1943): Bull. Math. Biophys. 5, 115-133

Mead, C., Mahowald, M.A. (1988): Neural Networks 1, 91-97

Mezard, M., Parisi, G., Virasoro, M.A. (1987): Spin glass theory and beyond. Singapore: World Scientific

Michalski, R.S., Carbonell, J.G., Mitchell, T.M. (eds.) (1984): Machine learning. An artificial intelligence approach. Heidelberg: Springer

Michalski, R.S., Carbonell, J.G., Mitchell, T.M. (eds.) (1986): Machine learning. An artificial intelligence approach, vol.II. Heidelberg: Springer

Miller, J.H., Forrest, S. (1989): The dynamical behaviour of classifier systems. In: Schaffer, J.D. (ed.): Proc. III International Conference on Genetic Algorithms. Los Altos: Morgan Kaufmann, pp. 304-310

Minorski, N. (1974): Nonlinear oscillations. Robert Krieger Publ. Co.

Minsky, M.L., Papert, S.A. (1988): Perceprons (expanded edition). Cambridge, MA: MIT Press

Minsky, M.L. (1986): The society of mind. New York: Simon and Schuster

Muehlenbein, H. (1989): Genetic algorithms and parallel computers. In: Pfeifer, R. (1989)

Muehlenbein, H., Gorges-Schleuter, M., Kramer, O. (1988): Evolution algorithms in combinatorial optimization. Preprint, to appear in "Parallel Computing". Amsterdam: North-Holland

Nicolis, G., Prigogine, I. (1977): Self-organization in non equilibrium systems. New York: John Wiley

Nicolis, J.S. (1986): Rep. Progr. Phys. 49, 1109-1196

Oi, T. (1987): Biol. Cybern. 57, 47-56

Packard, N.H. (1986): Complexity of growing patterns in cellular automata. In: Demongeot, J. et al. (eds.): Dynamical systems and cellular automata. London: Academic Press

Palm, G. (1982): Neural assemblies. Heidelberg: Springer

Parga, N., Virasoro, M.A. (1986): J. Physique 47, 1857-1864

Parisi, D. (1989): Intervista sulle reti neurali. Bologna: Il Mulino

Parisi, G. (1986a): J. Phys. A 19, L617-620

Parisi, G. (1986b): J. Phys. A 19, L675-680

Patarnello, S., Carnevali, P. (1987): Europhys. Lett. 4, 503-508

Peretto, P. (1984): Biol. Cybern. 50, 51-62

Personnaz, I., Guyon, I., Dreyfus, G. (1985): J. Physique Lett. 46, 359-365

Personnaz, I., Guyon, I., Dreyfus, G., Toulouse, G. (1986): J. Stat. Phys. 43, 411-422

Personnaz, I., Guyon, I., Dreyfus, G. (1986): Phys. Rev. A 34, 4217-4228

Pfeifer, R., Schreter, Z., Fogelman Soulié, F., Steels, L. (eds.) (1989): Connectionism in perspective. Amsterdam: Elsevier

Poundstone, W. (1985): The recursive universe. Chicago: Contemporary Books

Pribram, K.H. (1971): Languages of the brain. New York: Brandon House

Psaltis, D., Park, C.H., Hong, J. (1988): Neural Networks 1, 149-163

Rammal, R., Toulouse, G., Virasoro, M.A. (1986): Rev. Mod. Phys. 58, 765-788

Recce, M., Treleaven, P.C. (1988), in: Eckmiller, R., v.d.Malsburg, C. (1988), pp. 487-496

Riolo, R. (1986): CFS-C: A package of domain independent subroutines for implementing classifier systems in arbitrary, user-defined environments. Internal report, Logic of Computers Group, University of Michigan

Riolo, R. (1987): Bucket brigade performance: default hierarchies. In: Grefenstette, J.J. (1987), pp. 196-201

Riolo, R. (1988): Empirical studies of default hierarchies and sequences of rules in learning classifier systems. Ph.D. Thesis, University of Michigan

Robertson, G.G. (1987): Parallel implementation of genetic algorithms in a classifier system. In: Davies, L. (ed.): Genetic algorithms and simulated annealing. Los Altos, CA: Morgan Kaufmann, pp. 129-140

Robertson, G.G., Riolo, R. (1988): Machine Learning 3, 139-159

Rochester, N., Holland, J.H., Haibt, L.H., Duda, W.L. (1956): IRE Trans. Information Theory 2, 80-93. Reprinted in: Anderson, J.A., Rosenfeld, E. (1988)

Rosenblatt, F. (1958): Psychol. Rev. 65, 386-408

Rosenblatt, F. (1962): Principles of Neurodynamics. New York: Spartan Books

Rumelhart, D.E., McClelland, J.L. (1982): Psychol. Rev. 89, 60-94

Rumelhart, D.E., Hinton, G.E., Williams, R.J. (1986): Nature 323

Rumelhart, D.E., Hinton, G.E., Williams, R.J. (1986): Learning Internal Representation by Error Propagation. In: McClelland, J.L. et al. (1986)

Rumelhart, D.E., Zipser, D. (1986): D.Feature Discovery by Competitive Learning. In: McClelland, J,L. et al. (1986)

Schaffer, D.J., Grefenstette, J.J. (1985): Multi-objective learning via genetic algorithms. Proceedings IJCAI-85, 593-595

Schaffer, D.J., Grefenstette, J.J. (1985): Multi-objective learning via genetic algorithms. In: Proc. IJCAI-85, 593-595

Schaffer, D.J. (1987), in: Davis, L. (ed.) (1987), pp. 89-103

Schuster, H.G. (1984): Deterministic chaos. Weinheim: Physik-Verlag.

Schwartz, J.T. (1986): The limits of artificial intelligence. Technical report #212, New York University

Sejnowski, T.J., Kienker, P.K., Hinton, G.E. (1986): Physica 22 D, 260-275

Serra, R., Zanarini, G., Andretta, M., Compiani, M. (1986): Introduction to the physics of complex systems. Oxford: Pergamon Books

Serra, R., Zanarini, G. (1986): Tra ordine e caos. Bologna: Clueb

Serra, R., Zanarini, G., Fasano, F. (1987), in: Proceedings Cognitiva 85. Paris: Cesta

Serra, R., Zanarini, G., Fasano, F. (1988a): J. Molec. Liquids, 39, 207-231

Serra, R., Zanarini, G., Fasano, F. (1988b), in: Livi, R. et al. (eds.): Chaos and complexity. Singapore: World Scientific, pp. 230-234

Serra, R. (1988): Dynamical systems and expert systems. In: Pfeifer, R. (1989)

Sejnowski, T.J., Rosenberg, C.M. (1986), in: Anderson, J.A., Rosenfeld, E. (1988), pp. 663-672

Sejnowski, T.J., Rosenberg, C.M. (1987): Complex Systems 1, 145-168

Simonini, P. (1989): Proprietá dinamiche di sistemi cognitivi artificiali. Doctoral dissertation, Physics Dept., University of Bologna

Sivilotti, M.A., Emerling, M.R., Mead, C.A. (1986), in: Denker, J.S. (1986), pp. 408-413

Sivilotti, M.A., Mahowald, M.A., Mead, C.A. (1987), in: Anderson, J.A., Rosenfeld, E. (1988), pp. 703-711

Skarda, C.A., Freeman, W.J. (1987): Behav. and Brain Sci. *10*, 161-195

Smith, S.F. (1985): Flexible learning of problem solving heuristics through adaptive search. In: Proc. IJCAI-*85*, 422-425

Smolensky, P. (1988): On the proper treatment of connectionism. Behavioural and Brain Sciences *11*, 1-23

Sompolinsky, H., Kanter, I. (1986): Phys. Rev. Lett. *57*, 2861-2864

Sompolinsky, H., Crisanti, A., Sommers, H.J. (1988): Phys. Rev. Lett. *61*, 259-262

Steels, L. (1988): Steps towards common sense. In: Kodratoff, Y. (ed.) Proceedings of ECAI 88. London: Pitman Publishing, pp. 49-54

Steels, L. (1987): Self-organization through selection. In: Proceedings of the IEEE first International Conference on Neural Networks, San Diego, CA, vol.II, pp. 55-62

Steinbuch, K. (1961): Kybernetik *1*, 36-45

Suddarth, S.C., Sutton, S.A., Holden, A.D.C. (1988): A symbolic neural method for solving control problems. In: Proceedings of the IEEE International Conference on Neural Networks, San Diego, CA, vol.I, pp. 516-523

Surkan, A.J. (1988): Application of neural networks to the classification of binary profiles from individual interviews. In: Proceedings of the IEEE International Conference on Neural Networks, San Diego, CA, vol.II, pp. 467-472

Tesauro, L., Janssens, B. (1988): Scaling relationshpis in back-propagation learning. Complex Systems *2*, 39-44

Toulouse, G., Dehaene, S., Changeux, J.P. (1986): Spin glass model of learning by selection. Proceedings of the National Academy of Sciences USA *83*, 1695-1698

Touretzky, D., Hinton, G., Sejnowski, T. (eds.) (1988): Proceedings of the 1988 Connectionist models summer school. Los Altos: Morgan Kaufmann

Treleaven, P., Vanneschi, M. (eds.) (1987): Future parallel computers. Heidelberg: Springer

Treleaven, P. (1988): Neural computing. In: Proc. of the Conf. "Elaborazione parallela". Milano: Aica-Aei, pp. 319-331

Varela, F. (1984): Principles of biological autonomy. New York: North-Holland

Varela, F. (1986): The science and technology of cognition. Internal Report. London: Royal Dutch Shell

Vastano, J.A., Kostelich, E.J. (1986): Determining Lyapunov Exponents from Experimental Data. In: Mayer-Kress, G. (ed.): Dimensions and entropies in chaotic systems. Heidelberg: Springer

Weisbuch, G., Fogelman Soulié, F. (1985): J. Physique Lett. *46*, 623-630

Werbos, P.J. (1982): Application of advances in nonlinear sensitivity analysis. In: Drenick, R., Kozin, F. (eds.): Systems modeling and optimisation. Heidelberg: Springer

Widrow, B., Hoff, M.E. (1960): IRE Wescon Convention Record *4*, 96-104

Willshaw, D.J., Buneman, O.P., Longuet-Higgins, H.C. (1969): Nature *222*, 960-962

Winston, P.H. (1984): Artificial intelligence. Reading: Addison-Wesley

Wolfram, S. (1984): Physica *10* D, 1

Wolfram, S. (1986): Theory and applications of cellular automata. Singapore: World Scientific

Zhou, H.H. (1988): A prototype of long lived rule-based learning system. In: Proceedings of Computational Intelligence 88. Milano: ACM Italian Chapter, pp. 79-88

Subject Index